Footloose and Curious

Henry H. Hamilton III

To Clive, the story that won me a radio story telling stint on Public Radio for 7 years, is found here as #50. May you enjoy it and the others.

HHamilton

PRGott Books Publishing

Norway, Maine

www.prgottbooks.net

Footloose and Curious

Henry H. Hamilton III

Cover background map from *123RF.com*
Photo of the author on front cover and
photos on pages 3 and 205 belong to the author.

Layout by Laura Ashton
laura@gitflorida.com

First Printing

ISBN: 978-1725942639

Printed in the United States of America

PRGott Books Publishing

Dedication

In the 60s Barbra Streisand recorded "Never Will I Marry," with lyrics that stuck with me..."Wide my world narrow my bed." I associated this song with my dream to live an international lifestyle. This song indicated that I'd be doing it alone.

Then I met Shirley, who grew up on a subsistence farm with 10 siblings and a largely disabled father. That experience toughened her for the lifestyle I wanted. Confident that she could adjust to a life in mud huts and grass roofs, toilets—a hole in the ground—showers with the plastic pail and cup, running water with four legs, and all the rest of it.

We married and I found that with the right woman I could have both a wide world and a wide bed. She brought untold richness to the experience. She's life's greatest gift to me. I dedicate this book to her with all my love and appreciation.

Table of Contents

Table of Contents, Cont.

Table of Contents, Cont.

Table of Contents, Cont.

1. First Impressions

I was hired from Niger, Africa, by the United Nations to serve in Croatia. No one met me at the Zagreb airport, as had been promised. I had no contact name, address or phone number, only a job description and a project title in the village of Pakrac, somewhere in Croatia.

As I stood in a daze of exhaustion, unable to think, a big, friendly, English-speaking taxi man approached me. "Pakrac is not easy to get to," he said, "you might get there in 6 to 8 hours." Then, for an acceptable price, he offered to have me there in 2 hours.

With visions of a bath and bed only 2 hours away, I climbed into Mario's taxi and soon we were passing through a tree-covered autumn countryside of rolling hills, punctuated by newly harvested corn fields and quaint little farms of brick houses, sheds, and barns with red tiled roofs.

We followed a highway south to Novska where we turned east onto a narrow road, winding through thick hardwoods, their branches reaching over the road, nearly closing us in —a tunnel of fall colors of yellow and orange. The terrain became hillier, reminding me of drives through my native New England.

Mario's taxi was filled with the sound of Italian opera, complementing the lovely scenery.

Then we broke out of the woods into open farmland.

But the fields and gardens were overgrown with shrubs and bushes, indicating long neglect. The farm houses and buildings coming into view were windowless, the doors were missing. As we got closer I could see great holes in the roofs and the brick and stucco facades were riddled with hundreds of holes, large and small.

My first impression was that I was viewing the site of an isolated skirmish. But as we continued along the country lane, more and more farms came into view, all in ruins.

I was to learn that the road east from Novska, all the way to Vukovar, four hours beyond, represented the line of confrontation between Serb and Croate—it was a line of destruction.

The horrible recent history of this place began to weigh on me as we passed hundreds of blasted abandoned homes, representing the lives of thousands of people. I had visions of them under siege, herded like animals, executed, raped, abused. And these were not soldiers, but simple farmers, attacked by one side or the other because of their ethnic identity alone.

My stomach knotted and the Italian operatic strains, which had so perfectly fit my first impressions, now seemed to mock what I was seeing.

At last we saw Pakrac's sister town of Lipik in the valley below. From the distance it looked untouched. I could see a few people moving about, smoke rose from some chimneys. But from a closer perspective, here too, intense house-to-house battles, rocket attacks and artillery bombardment had left no structure untouched.

We passed through a vast complex of greenhouses, covering over 12 acres. I could see no panes of glass in place here, where once hundreds of people had worked to produce great quantities of flowers for Italian and German flower shops. Now only weeds sprouted from the glass-strewn ground.

What could I do to help these capable people? I remembered wondering when I first saw the lovely farms and perfect fields outside of Zargreb. But within an hour I had entered this zone, representing one third of Croatia, where that same beautiful world had been shattered—animals, crops and tens of thousands of people gone. Only empty shells remained, skeletons of people's dreams.

Will the dreamers ever be able to dream again? Is it possible to put together enough of the pieces that life here can again meet the potential of the place? Can the people ever overcome the rage and the terror over what neighbors can do to one another?

Well, that's what I'm here to find out.

2. Funeral

The supervisor of our roof repair team stepped on a spot of ice and fell twenty-five feet to his death. That was my second week in Croatia.

All staff members of my office attended his funeral. The experience was like something out of a Bergman movie. First we went to his mother's farm. That was, for me, like stepping back to my great-grandfather's farm. The tools in use here now resemble those only lying around the farms of my childhood, leftovers from previous generations.

The corpse was displayed in the family living room. The squatty old mother sat at her son's head, bemoaning his loss—her voice cried but her eyes did not. People brought flowers into the room, laid them in place, stood for a moment in silent meditation, and left.

Outside, the snow had begun to fall thickly and to gather. People sought refuge under porches, sheds and umbrellas. A dog barked at us from behind a shed door, and beady-eyed pigs oinked with curiosity. The barnyard was mucky from all the foot traffic. A snow drenched rooster dashed about snatching up worms exposed by the confusion of footprints in the mud.

At precisely two o'clock the casket was brought out and put onto a rickety old hearse wagon, which was attached to a tractor, not horses. Flowers were lugged out and piled on, the mother moaned, a priest chanted. Someone hand-cranked the tractor's engine, it backfired, caught on, and started to slowly pull out of the yard and onto the road toward town. The mother, supported under each arm, followed, and the rest of us folded into the funeral train behind.

It was snowing heavily; large wet flakes stuck to the woolen clothes and built up on umbrellas. Church bells pealed from the distance. At the edge of the tiny hamlet we turned onto a dirt road leading between two fields of corn stubble. We trudged up over a hill and found a graveyard and small chapel. As we entered the cemetery the bells from the village were replaced by a wailing siren. When it stopped, the casket was ready to be lowered by ropes into the ground. A young, well-fed priest, his black robes accentuating his fleshy pink face, said his spiel.

The old mother ranted and struggled to free herself from the two men who held her, as if to throw herself into the grave. Three bright yellow cars arrived. Eight men got out of two of them, toting their shiny brass instruments. With their music sheets wrapped in plastic against the wet snow, they *um-pa-paa-ed* a few sad tunes, as the two men in black ushered the old mother into the third yellow car and drove her away.

I joined the rest who trailed back to the farm with wet feet, chilled to the bone, thinking, here I am, in a country I had never considered being in until a few weeks ago, tramping through mud and snow, following the corpse of a man I've never met, taking part in one of life's dramas within cultural parameters unfamiliar to me.

How gloriously unexpected and various life can be.

3. Bathing Techniques for the Croatian Winter, 1996-97

Adjusting to the Croatian climate should be a snap for a native of Maine but this one has been softened by a decade in the tropics, and even in Maine we had central heating.

The frozen glass of water on my bedside table is tolerable as long as my body feels snug under a mountain of quilts. But it is bathing in such an environment that is my greatest source of discomfort.

I've developed some elaborate techniques for lessening the pain of what we usually consider a pleasurable activity.

First, I put my underwear and towel on an electric heater in the living room, the only source of heat in the house. In this way I am able to store heat in undies, socks and towels.

Take my towel for example. I lay it over the heater. When it is really hot I fold it, making sure the hot side is folded into itself. Then I lay the folded towel over the heater and repeat the process until I have a neat little package of heat. I do the same with my BVDs and long johns, as well as my socks.

Once they are all heated up I make a pile of them, towel on top, and stack them on the washing machine, which stands beside the shower stall. Though the exposed edges cool down quickly, in the center of the pile of fabric the heat is held in place.

The Croat part of town controls the water supply, but I live on the Serb side. Water is infrequently fed to the Serbs, usually in the middle of the night.

To cope with this, I open a faucet of the bathtub when I go to bed for the night. When, in the wee hours the water arrives it is accompanied by loud popping that awakens me. I jump out of bed and start collecting water in pails and jugs, while the on-demand water heater to the bathtub warms up bath water.

Now I turn on the hot water to the tub and heat up the glacial middle of the tub. (I have this shower nozzle on a tubular attachment which is designed to be hand-held). At a certain point the hot water heats up the metal faster than the metal cools down the hot water. Then the steam starts to rise. Oh boy, heat!

I strip off my sleeping clothes, sweat suit, long johns, undies and wool socks, and step into the tub, spraying with the steaming water whichever part of my anatomy is coldest. At this point I put the plug into the drain to keep in the bottom of the tub the water where it can release up its heat.

While one hand soaps up my head and face the other directs the spray of hot water over the body. Once the head is rinsed the spray warms the front while the back is soaped up and while the backside is being rinsed off I soap up the front.

Once completely rinsed of the soap I quickly rub off all the water I can with my hands and reach for the gloriously warm towel. I dry off, as best I can, while still standing in about 4 inches of warm water which continues to emit a little warmth.

By the time I am dry to my knees the water in the tub has lost most of its heat. I pull the plug with my toes, wipe one foot, step out and wipe the other. Then with great joy I put on the wonderfully warm undies, socks and long johns. Then I pull on a turtleneck sweater, a long shirt, pants and finally a V-neck sweater.

Now I am ready for the day.

These techniques work sufficiently in temperatures of 25 degrees or so, but in a few weeks it will be much colder. What shall I do?

According to Maine author C.A. Stevens' account of early 19th century settlers to Northern New England, perhaps due to the lack of buttons, the womenfolk would sew the men and boys into their long johns for the winter. This gives an indication of how much bathing our ancestors did during the frigid months and may indicate an option shortly open to me. Perhaps I should consider investing in a case of cologne.

4. *Mr. Zhili*

Visiting potential recipients of pigs from our United Nations Development Project animal distribution, my translator and I encountered Mr. Zhili, a tall, thin gentleman whose deeply wrinkled face was overgrown with an assortment of wens and other such epidermal growths of old age, giving his chiseled features the look of a lichen-encrusted ledge. His left hand was missing most of the fingers, but two short stubs remained just long enough to hold an ever-present cigarette.

"How can this decrepit old man, and his equally feeble wife, care for a pregnant sow," I asked my interpreter. Mr. Zhili hobbled along on his cane to open the shed door. There he proudly displayed three half-grown pigs. Their sty was clean and the pigs were well fed and happy.

I still had doubts. I am distributing expensive purebred Swedish Landrace pregnant sows that have to be repaid in kind to a neighbor. It is important for the full potential of the project that the pigs be distributed to capable people to assure that there will be a second-generation animal to pass on.

We left Mr. Zhili's little farm feeling that he could not be a first-generation recipient.

As we started to drive off, Franya, proprietor of the farm across the street, hailed us. He had been one of the very first recipients of this Swedish-funded pregnant sow distribution project.

"The best female from my sow's litter has been identified by the Government Selection Service," said Franya. "I've impregnated her according to our agreement and I am ready to repay my obligation."

"Do you think your neighbor Mr. Zhili is capable of raising a sow and her piglets?" I asked. Franya thought he was, and he pointed out that his elderly neighbors were very poor and that no aid organization had helped them at all.

The next day we returned to Mr. Zhili's to openly discuss with him our concerns about his ability to do the physical labor demanded by a sow and her litter. He insisted that the three pigs in the shed demonstrated that he was capable.

"Very well," I said, "your neighbor Fraya has a pregnant project sow that is repayment for the one we gave him last May. We will have him re-pay it to you."

"Oh no," said Mr. Zhili, "I don't want his pig."

We assured him that the sow was a purebred, inseminated under the Selection Service Supervision. He still was not interested.

"Do you have a problem with your neighbor?" I asked.

Mr. Zhili related to us a dispute he had had with his neighbor at the outbreak of the war in 1991. "We don't speak," he said.

"But," I pressed, "your neighbor spoke favorably of you when we discussed our doubts about your ability to handle a pregnant sow. We are here because he said you could do it."

The old man's eyes welled up with tears of emotion.

"Well," he said at last, "I haven't talk to my neighbor for seven years, perhaps we can start talking about this pig."

The deal was set.

Mr. Zhili walked us to his garden gate. As we reached our vehicle, Franya appeared on the other side of the road. Mr. Zhili did an abrupt about-face and hobbled back toward his house.

We told Franya that we'd agreed that his pig would be transferred to Mr. Zhili on Wednesday next.

As we drove off we saw Franya call to Mr. Zhili, who was about to enter his doorway, Franya motioned to the old man to join him at the gate.

The last glimpse we had of the two neighbors showed them to be talking together for the first time in seven years

5. Ben Bishop

Ben Bishop was a logger from Caratunk, on the Kennebec in Northern Maine. He took a job for the Great Northern Paper Company at their Jackman Camp, where he met a young Canadian bookkeeper, Felicity LeBout. When Benjamin Bishop's eyes fell upon Felicity he found himself in love.

Felicity LeBout was not flattered by the attention of men, but Ben Bishop's passion was true and persistent, so, in the middle of winter there was a wedding at Jackman Lumber Camp. The camp boss, Wade LaBeck, had a cabin on She-Bear Mountain and offered it to Ben.

Ben Bishop took his bride there with a basket of treats from the camp. They passed the evening with food and drink before the blazing fire, and then when Ben had taken his fill of wine he took his bride to bed.

With flowing flannel gown and raven hair,
a spirit she seemed to be,
but to his touch unyielding,
like a log in the bed laid she.

Ben Bishop laughed aloud, "Ah ha," said he,
"I see you wear a sullen frown,
but we'll change all of that;
take off your flannel gown."

Young Felicity slipped from out the bed
and straight for the door made she,
and flinging it back into the night,
bare footed and gown did flee.

Ben Bishop, astonished and muttering,
put on layers of wool and boots,
and out in the frozen night air
he hearkened to snowy owl hoots.

The moon was full and high
and lit the land like day.
Through the open wood, up over the ridge,
Ben Bishop followed her way.

Beside the path he spied her gown,
but much to Ben's surprise,
in the place of his wife's dainty nude foot,
the print of a beast he found.

Ben Bishop followed a little more,
seeing by moonlight drear,
'til the wood grew thick and the night turned dark,
Ben Bishop stopped in fear.

There from a wall of pine boughs black,
a glittering not from snow,
then again evil eyes fixed on his,
reflecting the moon's eerie glow.

With curdled blood and back of chills,

Ben Bishop turned and ran.
Pursuer now pursued,
a beast did follow man.

Through the wood they surged,
she closing in, caught Bishop by the door,
with fangs and claws and monster might,
limb from limb she tore.

Long stood ajar that cabin door,
and long the fire was out,
for never again in the world of men
was Ben Bishop seen about.

The pines still grow on that mountain side,
where no logger from Maine will crew,
for they say it's the lair of a she-bear,
who goes by the name LeBout.

6. Bouba Gets a Job

Remember Frances Gary Powers, the pilot who so embarrassed the Eisenhower administration when his U-2 spy plane crashed deep in the USSR? He flew from Peshawar, Pakistan, where the US Air Force had an intelligence base during that time. When I was assigned there in 1966, the installation was still active. It was located about 8 miles outside of Peshawar in the middle of a sugar cane plantation.

I came to know a certain middle aged Pakistani named Bouba.

Bouba had been unemployed for a number of years when he landed a job as waiter in the officer's club at the spy base, known as Budabare Station.

Suddenly Bouba had money, more money than he had ever imagined he could earn. The natives of this region are Pathan tribesmen, a people known for strong family ties and generosity. But Bouba had suffered so many years of poverty that he forgot the people who had been supporting him during the years of his unemployment.

Bouba wanted to enjoy the good life. He soon abandoned the baggy pants and long shirt of his people for a tailored three-piece European-style suit. His comfortable sandals were replaced by tight English shoes and stockings.

Soon he was riding a beautiful maroon bicycle imported from China, the Cadillac of the Frontier Provence, and lastly, a heavy golden chain adorned his rather prominent belly. Though it was rumored that the chain held no watch, it certainly made an impression. Bouba looked more like a banker going to work than a waiter.

Apparently this was also the impression of some of the sugarcane workers he encountered each day.

One evening, shortly after Bouba passed through the gate which guarded the American Base, the rotund waiter was seen staggering back along the road to Budabare Station, bruised, bloodied, and completely naked. Even his much coveted Fruit-of-the-Looms had been stolen.

Bouba didn't get much sympathy from family or friends. His selfish self-centeredness had received a just reward from the point of view of his people.

When Bouba's bruises had healed sufficiently, he returned to his job for the Americans. He rode his old bicycle and wore baggy traditional pants and long shirt as well as his old flip-flop sandals.

The next time I saw Bouba he was in the Kisakharni Bazaar buying shoes for his nephews and school supplies for his cousins. "It is written," said Bouba, smiling serenely, "that he who hordes has nothing, but he who gives has everything."

7. A Bird of a Different Feather

When I was a boy growing up in rural Maine special projects at my grandfather's farm were always exciting events for me. My family was a large one, and most members lived within an hour's drive of the family homestead. At grandfather's request, everyone would gather at the farm on a weekend to undertake one of these projects. Usually, the women would talk, prepare food, and watch children while the men worked. We children got to watch.

One such yearly project was winterizing the farm house. The clan gathered in October before deer hunting season, because once deer hunting season began the men would do nothing but hunt. Winterizing the farm house involved putting on storm windows and cutting pine boughs to pile up around the foundation. The farmhouse was set on a fieldstone foundation and the northern wind coming across the fields would blow through the cracks in the walls unless the pine boughs were placed there to prevent it. Of course, in the spring we had to reverse the process. Removing storm windows for storage was not very interesting, but what to do about the boughs? They were stacked into a huge pile in the pasture and set ablaze. That was exciting.

One such project that I remember with special clarity was the painting of the barn roof. Grandfather thought it was looking a little rusty, so we all gathered together to remedy the situation.

The men got up on the roof with sandpaper to clean up the rusty spots, and then they started to spread on the beautiful silver paint which glistened in the sun and was a marvel to look at.

Out in back of the farm, on the northern side, was a row of pine trees which had been closely planted together to serve as a wind break. On the southern side were huge hardwood trees which shaded the house and yard in the summer and let through the low sun in the winter.

For some reason, the local crows enjoyed gathering in the pine windbreak behind the farm. They would approach across the southern pasture and would group together in tight formation when they neared the barn, because the barn caused a break in the wall of hard wood trees through which the crows would funnel. They would skim over the barn roof and alight in the pine tree "stadium" for some sort of meeting or convention on a regular basis.

On this day they were so busy discussing their affairs that they did not notice the men on the barn roof. They skimmed over it at the usual altitude, allowing my Uncle Bob to snatch one of their party out of the air by the leg. Since he had a large paint brush dripping with silver paint in the other hand, he spontaneously painted the crow and let it free. The sun glistened on its wings as it flew off to join its pals.

The fellow crows were quite distressed when Uncle Bob grabbed one of their comrades; they protested loudly from the pines. But when their brother crow was released and tried to join them, their reaction was astonishing. Instead of welcoming their brother with his new glistening coat, they all flew straight up from the windbreak in terror. Then they gathered in a bunch and took off after their kinsman. They chased and harassed him mercilessly, eventually attacking him and driving him from their society. They did not find his new feathers as beautiful as I did, but different, too different to be tolerated.

What I witnessed made a profound impression on me. I was quite distressed by the mentality of crows. In the time since, I've noticed that most creatures, including—if not especially—human beings, quite often turn against their fellows who are different.

Were the crows behaving like humans I wonder, or do we humans manifest this birdbrain mentality? Our superior skull capacity at times appears to hold no more wisdom than a crow's.

8. Deacon Pike

The little country church I grew up in had a minuscule congregation. The boys and staff from a nearby summer camp would fill it for a few weeks every summer, but after they left for their city homes, the usual Sunday crowd rarely numbered over ten.

Such a small church could not support a minister, so we had what I guess you'd call an itinerant minister, who served several churches. He'd get up early Sunday morning and give a sermon from eight to nine in one church, and then he'd travel to our town and give a repeat performance from ten to eleven, followed by a noon sermon elsewhere still. This way, all the little churches in the area were able to share the burden of maintaining one, poverty-stricken minister.

Deacon Sylvester Pike, the most memorable character from that congregation, had a remarkable resemblance to the farmer in Grant Wood's *American Gothic.* He was a very long and lean somber person. I spent many a boring sermon studying his stoop shouldered stance and the great long hairless head which seemed to hang off his skinny neck. Surely he was meant to be a mortician. The gaunt deacon collected the offering each Sunday. He looked so stern and serious that I'm sure people reached deeper into their pockets than they might have otherwise, for fear of eliciting a frown from such a morose face.

Now, Sylvester was known for never saying anything mean or nasty or vulgar; he neither swore nor cussed, nor was he ever impolite. The men in town would often say of Sylvester, that he wouldn't say "dung" if he had a mouth full.

Of course, that was easier for Sylvester than for most folks, because the Deacon hardly ever said anything at all, and besides, his wife well made up for his taciturnity.

Deacon Pike eventually became aged, started ailing and took to bed. His wife tried all of the home remedies on him, but when he didn't improve she called the country doctor, who sent the Deacon to the hospital for some tests. The results were not good—terminal cancer. Deacon Pike was sent back home to die.

Doc Jenkins visited Sylvester after his return home. Doc told the Deacon that he didn't have long to live and that the time had come for him to make peace with his maker.

Sylvester lived but a short while longer, but during the greater part of that time, with all the energy and strength that was left in his body, he lay there on his death bed and he yelled and he screamed and he roared, all of those words, all of those insults and vulgarities, all of those abusive phrases that he'd never used all the days of his life.

His wife was so distressed by this turn of events that she would let no one into the house to visit Deacon Pike, explaining that he had gone crazy.

Well, after days of cussing and swearing and saying all manner of things that no one had ever dreamed that the Deacon Sylvester Pike even knew how to say, he quieted down. He slipped into a few days of peacefulness and serenity.

Neighbors and relatives were allowed to come in and say their last goodbyes, and then the good Deacon Pike gave up his ghost, as they say.

People never spoke openly about those things Sylvester said from his death bed, but the story was whispered in every home.

Some were quick to charge hypocrisy, but I'm convinced that what the Deacon said during the closing days of his life was his way of making peace with his creator, because Sylvester knew that he couldn't carry all of that stuff where he was going.

9. Dongo

During the 70s I delivered the town of Gardiner, Maine, for United Parcel Service. Gardiner was a bleak mill town along the Kennebec River. Edward Arlington Robinson grew up there, and in his poetry he referred to Gardner as Tilbury Town, a place where Miniver Cheevy, child of scorn, once wandered about, and old ladies killed cats with care.

My days in Gardiner were not happy ones. I was in a dead-end marriage and outer events seemed to be in charge of my life. It was not where I wanted it to be. My inner gloom must have shown on the outside. I didn't smile much, and this caught the attention of some of my customers.

There was a machine shop there where I delivered and picked up packages daily. The office girls were a highly spirited bunch. One day, a secretary asked me why I never smiled. The man who delivered rags always smiled and made the people around him happy. I should follow his example, she told me.

On one occasion I was at the factory when the rag man was delivering clean rags and picking up dirty ones. Sure enough, he smiled from ear to ear constantly. His tall lanky body, jerky movements, and insipid grin reminded me of Goofy, certainly not someone I wanted to emulate.

Then the secretary entered. "Have you met Dongo, our rag man?" she asked me.

How he could have smiled more I don't know, but his smile intensified. "He's always smiling so brightly, we wish you would smile some."

"I don't trust people who smile all the time, they're hiding something," I responded with an exaggerated frown. That comment certainly dampened the atmosphere, and I walked out of a leaden silence.

The next Monday that office had quite a different atmosphere. When I entered the people rushed up to me and asked if I'd seen the paper. I hadn't.

They held up the headlines: "MOVIE DIRECTOR MURDERED BY HIRED GUN, DONGO ARRESTED." There was a photo of the beloved Dongo in police custody, handcuffed, but smiling.

It was a sensational case, and although prosecution proved that Dongo had negotiated a contract to kill the movie director, his argument held that he had chickened out of the deal and that another hired killer had done the job in his place.

Dongo got off the rap, as they say, but fate caught up with him. It seems that the smiling ragman and possible hired gun was also a drug dealer who didn't treat his associates in crime too fairly. He didn't understand or chose not to heed the rules about honor amongst thieves. Three of them kidnapped Dongo and drove him to the lovely country town of Bridgton, took him out into the woods and executed him.

Dongo made the headlines again.

"See what smiling gets you," I told the girls in the factory, who had become sobered by the realization that they had been charmed by such a demonic character.

I would like those women in Gardiner to know that the circumstances of my life have changed greatly since then. I am back on top of things, I find myself where I want to be, and I can now smile with ease, but mind you, only for a good reason.

10. Gifford's Secret

Alice and Gifford were elderly neighbors during my childhood. They had no children of their own but were very close to my sister and I who referred to them as Aunt Alice and Uncle Giff.

I have fond memories of these two people. Every year Alice gave me a birthday card containing a brand new, never-before-folded, one dollar bill. She kept this up far into my adult life. I still have the last one she sent me while I was a staff sergeant in the Air Force.

Giff genuinely enjoyed children. He had a couple of favorite tricks especially for kids. One involved holding a kitten in his huge hands in such a way that it looked as if he had separated the kitten's head from its body. The head would stick out of his gigantic closed fist from between the thumb and forefinger, and the feet and tail would stick out here and there between other fingers. The sight was astonishing, a head so far removed from other body parts certainly appeared to be detached.

On Halloween Giff enjoyed visiting our house before we went out trick-or-treating to see our masks. We would gladly display them for his approval.

"You certainly look frightful," he would say, "but you know, old Giff doesn't need a mask to look scary."

Then he'd scrunch up his old rubbery face in a grotesque

expression and would open a cock-eyed grin, exposing his old gums that contained only about ten teeth. We children all agreed that Giff wouldn't need a mask to take the scary prize.

Now as Alice grew older she grew more and more fearful that she and Giff would run out of money. She took full control of the purse strings, and her husband had to bargain hard to get a dime.

Giff always bought old rattletraps of cars, would soon grow tired of them, and exchange them for another. However, with Alice now in charge of the cash flow, that wasn't so easy.

I remember a time when they were having quite a disagreement over the need for trading cars. As Giff was driving Alice to church one Sunday, they crossed a causeway where Giff suddenly lost control of the car. He swerved all over the road and nearly went into the lake.

The next day Alice handed over the money for a new old car.

Between the military and college I visited Alice and Giff. Giff complained to me bitterly that Alice had eliminated sweets from their budget. There had been no pies or cookies in the house for years, he told me.

Well Giff soon escaped these frustrations of life by passing on. Since I was in need of a car, I offered to buy Giff's old VW from Alice. She explained that she would have to ask at least $100 for it, because over the last year of Giff's life, she had had to replace the clutch, transmission, brakes and battery.

I eagerly paid the price and headed for college. I got less than two miles, however, when the clutch went. The next weekend I brought home a friend who specialized in VW repair. He assured me that the clutch, transmission, brakes and battery were all original equipment.

A used clutch put me back on the road for a few days, but in Maine, a car must pass a yearly inspection, and no one would pass this bucket of bolts. I sold it for junk. Old Giff's gouging his wife ended up getting me but I didn't lose much on the deal.

Of course I am sure that if I had returned the car to Alice, explaining the situation, that she would have refunded me my money, but I decided to keep it all a secret between Giff and myself.

I was very curious about how Giff might have used the money his wife had handed out for car repair.

Years later while discussing old Giff with some friends, I learned that during the last year of his life Giff was often seen at a cafe in the next town which specialized in homemade apple pie and chocolate chip cookies.

If I hadn't forgiven Giff before, I certainly did then.

11. Goliath

Many years ago, while I was serving in the Air Force in England, *Camelot*, staring Lawrence Harvey, was on the boards in London. A friend and I traveled into the city from Bedford to view the play.

We bought our tickets at the theater, and with three hours to kill before show time, we wandered around the Soho District, the center of London's night life. We found a bar with swinging doors reminiscent of an old western saloon. We went in, took a table by the window, and each ordered a pint of light and bitter.

The evening was early and the bar wasn't crowded. A few parties were scattered amongst the tables, and at the bar sat one man. He was small, around 5'4" and 120 pounds, not very healthy looking.

He sat at the center of the bar with his arms folded on the white marble top, his forehead resting upon them. In front of him sat a glass of scotch and by his right elbow stood a large lead crystal seltzer bottle. Now and then the *petit* fellow would raise his head, take a sip of scotch, and then rebury his face in his arms.

Suddenly the swinging doors slammed open and in strode a huge fellow, a block of a man. He approached the bar with a John Wayne swagger and loudly demanded a bottle of London Pride. He stood at the bar beside the little man with the buried head, placed both his huge hands on the marble top and bellowed that he could whoop any one man or any ten women in the place.

Nervous laughter swept the room because no one doubted his claim. He loudly boasted that he could throw a man through the window, clear across the street, and through the window of the sandwich shop on the other side.

How serious he was in his threats, I'm not certain. I tend to believe that his claimed, though obviously true, physical abilities were in jest. He was playing with us. The little fellow beside him seemed to take offense. He turned his head a bit and looked up at the boaster with one eye. Then he grabbed the heavy seltzer bottle and smashed it on the giant's hand. The giant screamed out in pain and held up a bloody hand with mangled fingers. The swinging doors slammed and swung, and the little man was there no more. My friend and I decided to follow his example.

Once outside we made our way to Dreary Lane Theater, where we thought it might be safer to wait for curtain time. Though the spectacular Camelot was the first professional play I'd ever seen, I remember with much more clarity the fall of Goliath to the little drunk in the bar, than I do the good king, Lawrence Harvey, and whatever it was that the simple folk did.

The scene in the bar was a hard act to follow.

12. Horris Tillson

The New England character, they say, is as solid as the granite under the New England soil. It is marked by a unique and a subtle sense of humor, though New Englanders are also well known for taciturnity and stubbornness. Our Puritan heritage has left some of us more into guilt and blame than folks from other regions, I believe.

But by the same token, the Puritan influence has left many New Englanders with a profound appreciation for hard work, frugality, and the democratic process. If you don't believe it, try attending a New England town meeting when the town budget is being discussed.

One characteristic of the New Englander that has impressed me over the years is the "unflappableness" of the people here, and last week I met a prime example of this unflappable type.

I was traveling through a rural area in central Maine when I saw a sign at the end of a dirt road. "Tillson's Rabbitry" it read. Being an animal scientist with much experience in rabbit raising, I was eager to see a backwoods rabbitry in operation.

About a mile down the dirt road I came upon open fields and a small farmhouse and barn, with a long rabbit shed to one side.

Tillson's farm was beautifully manicured. With a John Deere riding mower, Horris Tillson kept clipped about three acres around his well-maintained buildings. Flower gardens were tastefully arranged; he kept a perfect little vegetable garden. The place was immaculate.

However, about 50 feet to the left of the house grew an unsightly clump of scraggly bushes in this otherwise park-like landscape. Blackberry bushes grew there, alders, birches, poplar, and a sumac or two. It looked quite out of place.

Mr. Tillson greeted me with open friendliness and enthusiastically showed me his rabbit project. He was in his late 70s, a widower. He raised rabbits to supplement his diet and income, but mostly he raised rabbits for the sheer pleasure of it.

His rabbit barn was clean and fresh smelling; his animals were bright eyed and healthy. We swapped tales about rabbit raising and agreed about the satisfaction derived from working with animals. It was a good visit.

As Mr. Tillson showed me to my car, my eye again caught sight of that ugly clump of scraggly bushes so close to the house. Then I noticed amongst the bushes a number of well tiles.

"I'm curious, Mr. Tillson," I said, motioning toward the bushes, "did you change your mind about having a dug well?"

"Nope," said Mr. Tillson, "I got a dug well out by the barn, there. But about fifteen years back me and the Mrs. we went out grocery shoppin' one evenin,' and when we got back, why, we found this bunch of four foot concrete well tiles layin' there, twelve of 'em. I'm just waitin' for whoever mis-delivered 'em here to come back and get 'em."

Horris Tillson, a personification of the unflappability of the New Englander character.

13. Ralphie

Do you hold any grudges? Most of us prefer to believe that we don't. But if we carefully search the darker corners of our inner terrain, there is sure to be some long forgotten resentment lurking there, some act or attitude, some social slight that offended our perception of ourselves. In truth, most of us carry a little grudge or two. I often hear evidence of it in the tone people use when describing long past events.

For example, years ago, Ralph and Dorothy Lamb, a very elderly couple from my home town, traveled to London where I was living.

We met for dinner one evening, and Ralph told me about being a schoolmate of my great-uncle Charlie. They had been great pals, he said, but then his eyes narrowed. He and Uncle had shared a ruler, he told me. Rulers in those days were veritable boards, solid oak, 18 inches long and 3 inches wide.

Because Ralph and Charlie tended to get into mischief when they sat by one another, the teacher had separated them to either side of the room. If Uncle had the ruler and Ralph wanted it, he would give a sign, and when the teacher's back was turned Uncle Charlie would toss the ruler over the heads of the other students to Ralph.

Once, when Ralph was in possession of the ruler, Uncle Charlie gave the sign, and as soon as the teacher turned to write on the board Ralph threw this plank across the room to its co-owner.

Now, Uncle Charlie sat in front of the windows, and when he saw this ruler coming a little higher than usual he chose to ignore it. It crashed through the window. The teacher demanded to know what had happened and who had done it. Several classmates were quick to say that Ralphie had thrown his ruler through the window. Ralphie got yanked outside by the ear and got paddled on the can with his end of the measuring stick.

Even though Ralph was telling me this story some 70 years after the fact, his old eyes still snapped with anger and indignation.

You see, he had been the recipient of what he perceived as unjust punishment. He had never worked it out with Uncle Charlie, and by that date Uncle Charlie was no longer of space and time.

So, if you're carrying around some little grudge like this, and I know you are, I suggest you resolve it while you can. Probably the object of your ire will have no memory of the incident that you remember so clearly, but if you can summon the courage to broach the subject, I am sure that the two of you will share a good laugh over it and then you can relegate it to the past. You may even reestablish a long-lost friendship.

I've tried it and I recommend it.

14. Remembering Dr. Greenleaf

From a kid's perspective Dr. Greenleaf was a stern old man. He looked like the farmer in *American Gothic*. He practiced dentistry before speed drills and used instead those head rattlers run by little cables and pulleys. Novocain was for teeth pulling not for drilling out cavities, and Dr. Greenleaf didn't respond kindly to complaints about pain. His wife assisted him and she countered his sternness with kind, gentle words and pats.

My father worked in the Oxford Woolen Mill and my mother was a homemaker. I now know that they had to greatly sacrifice to keep their children's teeth in shape. I appreciate it now but didn't then. Dr. Greenleaf was the most dreaded person of my childhood life.

Well, it seems that from conception our genes dictate the number of cavities we'll have. By age twelve, mine had all expressed themselves. Fortunate timing, because Dr. Greenleaf died about then.

Years passed and during my senior year in high school, a filling fell out of a molar.

Dad took me to the new dentist in town, Dr. Q. He had a state-of-the-arts set up—speed drills, freely used Novocain, and promised a no-pain experience. He checked me out and then called my dad into the room.

"You see these white lines along the base of the teeth?"

he asked my dad. "These will soon all turn into cavities and should be drilled out and filled now. In addition, all these other fillings will be falling out so I should replace them now. I can do this for $385 (1962 dollars) and you can have up to two years to pay for it."

I was committed to the military as soon as I graduated in a few months, so we decided to defer that expense to the US Government. Dr. Q replaced the one missing filling.

The next summer at USAF Basic Training at Lackland, Texas, a military dentist looked into my mouth. "Where did you get this work done?" he asked. "Hometown dentist, Dr. Greenleaf," I answered. The AF Captain called other dentists into the room to look at my dental work, the most magnificent dentistry he'd ever seen, he told the others.

It's been nearly 50 years since Dr. Q's prognosis. The white lines at the base of my teeth are still white, and the only filling that has dropped out was the one Dr. Q put in. Thank you, Dr. Greenleaf. Didn't like you as a kid, but admire you as an adult.

15. Stella

Stella Edwards, my 6th, 7th, and 8th grade school teacher, was a rotund lady, invariably dressed in grays and blacks. Something I couldn't articulate made her different from other people and teachers I knew. Later, the term I learned was *intellectual;* Stella had a marked intellectual orientation.

She would sometimes remind me that I had a large family that would be proud of me if I did my best. At this time in my life I was more influenced by my peers than by any stuffy adults, family or not. The influence of my peers, however, did not last. Stella's influence did.

When I was a graduate student at the University of Paris many years later, I wrote to Stella, telling her how she had been my role model to a large extent during my youth. Although my actions under her tutelage probably never indicated it, her values had been important in shaping my intellectual development and had something to do with the grand experience of my being a student at La Sorbonne.

A decade later in my life, on a visit to my hometown and parents, Mother told me that Mrs. Edwards, as we always called her, was in a nursing home nearby and had told my mother that she would like to see me.

From my visit to Stella I discovered that although she had suffered the amputations of both legs, her powerful spirit still reached out to influence her now diminished sphere. She organized social events at the nursing home and was in charge of the Welcoming Committee.

When I walked into her room she was sleeping. Her short, legless body looked mostly dead. Tubes led under the blanket and her once rotund body had wizened away.

I touched her hand, "Stella," I said.

Her eyes opened. "This is Henry Hamilton, I've come to visit you."

Her old body suddenly became animated; she pulled herself up into a sitting position and excitedly asked me to get her glasses out of her bedside table.

Her glasses were now so powerful that they magnified her soft blue eyes to inhuman proportions. She asked me to sit on the bed beside her so that she could get a good look at me. Then she took my face in her warm hands and pulled it to within six inches of her own, carefully looking my face over, portion by portion.

Then, still holding my face she softly said, "They tell me you've been through divorce. Did you find it terribly painful?"

I nearly wept. That this suffering moribund old woman would call me to her bed to comfort me about my divorce touched me deeply. It was a powerful indicator to me of her depth, breadth of spirit and character.

A few months later at her funeral, the officiating clergy said, "Whenever I went to the hospital or nursing home to comfort Stella, I always came away feeling that I was the one who got the most comfort."

I knew first-hand what he was talking about.

16. Sylvester

I helped out with the local 4-H club while living in rural Maine a few years back. I specialized in photography, forestry, woodworking, and outdoor cooking.

One summer the county agent asked me to participate in the annual 4-H Field Day. This is a county-wide event where all the local 4-H clubs gather for a day. Here 4-H'ers may choose to participate in various educational activities.

They can learn how to trim fruit trees, sex rabbits, prepare an animal for show, etc.

I organized the woodworking activity. To simplify the process and to minimize the number of tools required, I precut all the wood. The children had only to nail together a simple shadow box and stain it.

My table was so popular with the children that they had used up all my materials by ten o'clock. That was okay, because by 10:30 it began raining.

We were on the grounds of an academy, so all projects could be moved into the gymnasium away from the rain. I went to the gym and sat on the bleachers for a while, watching the pandemonium occasioned by the move. Soon I was occupied with one of my favorite pastimes, people watching.

I noticed a young man in the multitude who stood out for several reasons. He was somewhat formally dressed for a 4-H outing, he was large for his 12 or 13 years, and stoop shouldered. He shuffled amongst the many present with no apparent aim, alone in a crowd.

But the most outstanding characteristic of this young man was that he reeked of guilt. Whenever an elder looked at him he responded as one caught peeking through a key hole, yet I could not see that he was up to any mischief.

Perhaps when an adult looked at him, I speculated, he was sure that the secret facts of his youthful autoeroticism were clearly evident, his reaction had that sort of tone to it.

Then I recognized a familiar face, a local entertainer who had a leather crafts table. He seemed to be overwhelmed by the number of children around him, so I proceeded to his table to help.

Almost immediately a busty matron bustled up with this stoop-shouldered, guilty looking lad by the collar.

"I don't know what this character is up to," she spouted, "but I can assure you it's no good. I want him to stay here with you men so you can keep an eye on him." Then, after fixing him with an accusing eye, she disappeared as quickly as she had appeared.

"Welcome, my name is Henry," I said.

"I'm Sylvester," said the kid, looking at his shoes.

"We're making simple coin purses, knife sheaths, even wallets if you have the time." I told him.

"I'll make a coin purse," said Sylvester obediently.

He proceeded with his work, never looking up from the table, as if resigned to a punishment.

Before long a tall man announced that the rain had stopped and that chow was ready. We were to go the food tent, collect our lunch, and proceed to the ballfield where a special guest was about to give a rocketry demonstration. A murmur of excitement spread amongst the youngsters.

Sylvester abandoned his partially completed coin purse and disappeared into the crowd, while I beat a path for the hamburgers, potato salad, chips and punch. I found a tree to support my back by the ballfield, and with the kids from my club, settled down for lunch and blast-off.

Soon the county agent made a short speech, thanking all of us for participating, and then he introduced our honored guest, the first place winner of the state science fair competition.

It may not surprise you that this honored guest was none other than the stoop-shouldered, overweight, guilty Sylvester. He shuffled out onto the field with a reasonable degree of confidence for such a kid. He was accompanied by two half-pint, admiring assistants. They arranged Sylvester's hand crafted balsam-wood three-staged rocket and set it off with great precision.

I don't know what the busty matron thought of Sylvester's demonstration, but certainly everyone else was favorably impressed.

Sylvester was a hit.

17. Zulf

Decidedly middle-aged am I, but still in top physical condition. I mean, pushing 50 and still pumping a bike through the desert of SW Niger, through sand and dust and gravel and the blast-furnace wind, temperatures up to 120°. Perhaps the villagers do call me DO-TEE-JO (respected elder) and MA-GHIDA-BABA (Chief Father) but this is the stuff of youthful vigor!

It's exhilarating to speed through the bush with this flattering image in mind, terrified quail bursting out of the path, worried goats scattering before the sheer power of my movement, villagers gawking in disbelief as I fly by in gear 21.

Why, I can easily believe that I'm 21! At least I could—until I met Zulf.

Might it be, I began to wonder, that my many geared mountain bike permits certain levels of illusion? Is it my great shape and youthfulness, or is it the technology I'm on? A mere shift in gear ratios and I can take the power as my own.

Well, Zulf the fishmonger brought this kind of reality back into my life. Zulf peddled a wobbly-wheeled steel Nigerian bike, as basic a bike as you could find. Cord cut from an old inner tube held a jute sack of fish destined for the Torodi Market, strapped to his flimsy carrier.

The path up the plateau from my village is narrow,

and Zulf's one-speed rickety wonder was slower than I appreciated. I was impatient to get around him on my state-of-the-arts machine. An ironstone outcropping gave me the opportunity I needed. With a burst of power I pulled around the surprised Zulf and his smelly wet bag of fish and was free to double my speed.

Zulf took the challenge—he remained one bike length behind me. A flick of the lever and a little more power and I surged ahead still faster, but Zulf was right there.

Then we broke onto the top of the plateau. *Okay, Zulf*, I thought with zeal, *I'm about to give you a little lesson on gear ratios, and off I sped, pushing hard.*

But to my slight irritation Zulf remained in tandem.

Very well, I thought smugly, *let's see how long you can keep this up, you whippersnapper, you!*

On the plateau's decline I clicked into top gear and left Zulf behind—his legs simply couldn't go fast enough to keep up. But there as I mounted the next incline, he would be on my tail again, fairly pushing me along.

So, I see, I thought with resolve, *this will be a test of endurance. He has to exert himself many more times that I do, he can't keep this up all the way into town, I'm certain.*

The race became less fun now, more work—temples pounding, perspiration flowing, throat aching, I grimly pushed on.

At last Torodi came into view, its outskirts marked by a dry stream bed. If I hit the sand just right I knew that I could blast through and mount the clay bank on the other side. Zulf was still only a few feet behind me. If I failed we would crash.

I made it through safely and dug up the other bank several

yards ahead of Zulf. It was here, however, that I ended the race. I stopped in the shade of a spreading bani tree for a drink of water, as Zulf sped past.

Fumbling with my canteen, drenched in sweat, every cell pounding, hands shaking, I watched Zulf disappear amongst the jumble of adobe buildings, the real winner, I had to confess.

Yes, youthful vigor had triumphed over respected elder and his lightweight technological Western Marvel, gears and all.

I drank deeply the warm water from my canteen, thinking, DO-TEE-JO, MA-GHIDA-BABA, not such bad titles.

Chief Daddy, okay. I like that!

18. Fishing Lessons

"Junk fish," said Dad,
As he threw it off into the woods.
A shimmering shiner
lay in radiance
on the velvet green moss.
Thrown there to die.
Punishment
for taking trout bait.
The red and silver brilliance,
quivering
in the forest filtered light,
filled a child with wonder
love and sorrow,
as he secretly grieved
for the miraculous
reject.

19. Grandmother's Visitation

Last Saturday I experienced the distinct pleasure of a hot dog feast in the back yard of the Watters family. Kathryn Tucker-Windham was there, and toward dusk she told us some ghost stories.

In the aura of the rather spooky atmosphere created by Kathryn's stories, host Harry asked me if I had ever experienced ghosts. I told of a long past encounter, completely forgetting an on-going situation in my life right now.

I live in King Village, one of the last places you'd expect to encounter a spirit, but our apartment is so blessed.

I'm an avid photographer and on our bedroom walls alone hang over 20 framed photographs. Two of them are of my grandmother. One pictures her as a young woman of 18, the other is the last photograph I ever took of her at age 85.

I grew up across the road from my grandmother's farm, and she was as much a mother to me as was her daughter, my biological mother. She had time for children's activities. I felt privileged above all my cousins to live so close to her.

During her decline I became the only member of the family she could recognize or remember. My presence would sometimes bring her out of the depth of her senility and she would call me by name and speak coherently. I loved her completely.

Grandmother died about ten years ago and I've always

felt her spirit to be with me, but here in Huntsville she has made her presence known.

About nine months ago my wife and I heard an unusual sound in the night, a soft tapping. We turned on the lights to investigate. We found the photograph of Grandmother at age 85 rattling on the wall. Three other pictures hanging within inches of hers did not rattle.

We had been involved in amorous activities—we must have shaken the house to its very foundation, we reasoned.

The next time Grandmother rattled, no amorous activities had taken place, but the furnace was running, perhaps it caused the wall to vibrate.

The third time the furnace was not running, nor had any recent physical activity taken place in the house, however, the wind was blowing fiercely.

Then, as if to confound our intellectualizing about things beyond the intellect, the picture on the other wall, Grandmother at 18, began to rattle.

My wife and I smiled knowingly at one another accepting Grandmother's visit.

Grandmother still rattles away, on a regular basis, for five to fifteen minutes each time, from one wall or the other. My wife and I listen to her with the calm assurance that she is sending her approval of what we are doing with our lives.

It is so very fine to have Grandmother here with us in Alabama.

20. Gus

Though Judy (WLRH Sundial Hostess) has described me as a creative writer, I do not claim the title myself. I once signed up for a creative writing course in college, but when the professor demanded a composition on the subject of my right index fingernail, I filled out an add/drop card instead, dropping Creative Writing, and adding Greek Drama.

I write about personal experience, a subject that mostly creates itself. My function as a writer is to edit experience so that it conforms to the point I wish to make, or to the restrictions placed on me by the consumer of what I write. Those restrictions invariably demand a shorter format than I'm comfortable with.

One such experience involves a fellow I met while in Peace Corps training, Gus. He was going through some sort of crisis, perhaps mid-life.

Quite out of the blue one day, Gus handed highly critical letters to several Peace Corps administrators, threw away everything he owned that wouldn't fit into a daypack, and went stomping off never to be seen again by any of his fellow trainees.

It was an abrupt departure which shocked all who knew him.

Those of us who were his dorm mates looked at the trash cans filled with his rejections with a certain discomfort.

When we saw the photograph album in the trash we retrieved it and got a clearer view of the man's past and personality than he had ever divulged. Gus was very organized. His photo album was perfection. He had carefully selected and arranged photographs from the years of his life. The new military recruit was there, the mature soldier in Vietnam. He had traveled widely and in some very exotic places.

After leaving a career in the military he had grown a beard and had taught math to Eskimos in Alaska; it was all there, captured in precise definitive photographs—his life now thrown in the trash.

I had never felt close to Gus. He had an angry aura about him. He was basically a fundamentalist. A fundamentalist as an atheist, a vegetarian, jogger, a fundamentalist voice against pesticides, fertilizers, slash burning, and a voice for the wonders of composting and the defense of rain forests.

I could not quibble with the issues he stood for, but like his photographs, everything seemed to be black and white for Gus.

He was not an easy fellow to like.

His life in the trash, however, seemed to touch all of us. He was a fellow traveler in life who had reached some sort of dead end. I guess we felt sad and rejected that he had not shared any of it with us. Perhaps we felt that he had thrown us in the trash as well.

We removed all the photographs and threw the album back into the trash. We put them in a large envelope and saved them for three months. Then, hoping Gus was a bit more grounded, we asked the Peace Corps Director to send them to his permanent States-side address.

None of us ever heard from Gus. We don't know if he

received the photographs or appreciated them if he did. But we were doing as much as we could to reach out to another troubled human being, and at least we felt good about that.

21. Backwoods and Bankers

I traveled back home from my Peace Corps post in West Africa this summer to attend my daughter's wedding. My old college roommate, who now works for a multinational banking concern, traveled to the wedding from Europe with his whole family. He is my daughter's godfather.

He wrote me in Africa that he'd like to spend about two weeks, after the wedding, showing his family around the grand state of Maine.

I had a brainstorm. Knowing my limited budget, family and friends in Maine had offered me places to stay all over the state, on sea islands, on lakes and on a river. I proposed to my friend that my wife and I be his family's guides. We would arrange for shelter, and he would rent a vehicle big enough for all of us and would provide for food and fees.

It was a deal, and after the wedding my wife and I, my friend, his kids, pregnant wife, cousin, and a nanny all piled into a large van to tour my home state. We planned hikes, canoe trips, river rafting, a whale watch, fossil hunts, and food—wonderful food: lobsters, steamers, fried clams and blueberry pie. It was to be a grand celebration of the best that Maine has to offer.

We did have to deal with one slight problem, however. My friend was in the process of negotiating the sale of a two billion dollar oil refinery. He carried a satellite beeper which would go off at the most inconvenient times. Then we'd madly seek out one of those rare backwoods payphones, and he'd talk to New York, DC, Houston, London, Paris and Rome for hours.

One morning I dashed him off to the Bangor Airport for a private jet to JFK and the Concord to Rome. He was back for dinner after signing a document in Italy. These jet-set interruptions didn't spoil our fun, but they certainly scrambled our best laid plans.

We were breakfasting in The Forks when my friend got a message to get himself to where he could receive an 85-page fax. That would be Portland, two hours away.

We loaded up the entire family to head south, but along the way we passed a road leading to Houston Brook Falls, where the family had frolicked a few days before. The children pleaded to go there again.

My friend calculated that we could spend 45 minutes to an hour there, so I turned off route 201 and headed for the backcountry and the falls.

We parked off the road by a marked path, and as the last door of the van closed behind us, my heart sank. The keys were inside this newfangled rig with automatic locking doors!

My friend looked at me sternly. "You didn't," he remarked, as we stood there in our bathing suits and bare feet, miles from civilization, phones and fax machines.

"No problem," I assured him with mock confidence, a specter of catastrophe looming in my imagination. "You just go for a swim and I'll take care of this."

My guardian angel must have been watching, for no sooner had the gang disappeared into the woods, than a pickup rattled down the isolated road. The drunken driver drove me to Bingham and a phone booth. "The nearest Triple-A service is in Skowhegan, they'll take hours to get here," the drunk assured me. "After I pick my wife up from work I'll come back with a coat hanger and we'll get that thing open."

Triple A was faster than the drunk. In about 25 minutes a wrecker arrived and we headed back out into the woods. I told the young driver of my plight and of the helpful drunk who had given me a ride.

"His heart was in the right place," said the young man, sternly, "but his head wasn't. Only a fool drives drunk. Folks like that kill people and abuse their own families. Irresponsible, I'll tell ya. Why, I wouldn't touch that stuff for anything. I've got a wife who's a gift from God and two beautiful daughters. Booze would do nothing but ruin all our lives. I know, 'cause my father was an alcoholic."

He continued his lecture on the evils of alcohol until the van came into view.

Good news, my wife and friends were still in the woods playing in the water. The young teetotaler opened the door of the van in the blink of an eye. I grabbed the keys and dashed off through the woods to greet the happy gang of adventurers as they were picking up to leave this natural water park. My little deviation hadn't put a wrinkle in their day at all.

We sped off to Portland, my friend received his faxed mega-document, responded, and now a bunch of Texans are on their way to owning a European oil refinery.

Everyone had a satisfying experience and international big business proceeded, thanks in part to a big-hearted drunk and a conscientious abstainer from the backwoods of Maine.

22. Beans

Some years ago I found myself traveling through Southern Maine, headed for my home stomping grounds, when I spotted a sign on a church: Bean Supper Tonight, it read.

Memories of the community and church Saturday night bean suppers came flooding back to me. They always spread out a great variety of beans and no two pots tasted the same. My habit was to take one spoonful from each and every bean pot on the table. I preferred beans in a dark rich molasses sauce, but most of all I loved variations on the kidney bean.

Since my business that day took me within a few miles of my parent's home, I dropped off for a visit. I told them about the Bean Supper sign I had seen. I talked of my fond memories of such suppers and about how it had been at least a decade since I'd tasted any beans but those from a can.

My mother informed me that I was in luck because she had just put a pot of beans into the oven and that I should return in the evening to share supper with them.

I went about my business that day with visions of home baked beans in my head, so that when I returned to Mother's kitchen that evening my tongue was well primed for the taste of beans.

The problem was that the visions I had been having were of kidney beans. When my mother set a large plate of pea beans before me I couldn't hide my disappointment.

Of course, I thought, *that's all she ever fed us when I was a kid.*

I suddenly remembered that one of the advantages of those public Saturday night suppers was that I got out of my mother's beans. She always served those measly pea beans every Saturday night, and I found them a real chore to get down.

"What's the matter, don't you like my beans?" asked Mother.

"Not bad for pea beans," I answered, "but I had my mouth all primed for the taste of kidney beans and it's taking me a moment to make the adjustment."

"Kidney beans!" exclaimed Mother, "I'd never make those things. That's all my mother ever fed us kids when I was growing up, I can't stand them. When I left home I swore I'd never eat a kidney bean again, and I haven't."

Well, I could certainly sympathize with her attitude.

It's been ten years since that disappointment, and I still haven't tasted beans like those I remember from my childhood. I always keep my eyes open for a Bean Supper Tonight sign, and if I ever see one again you can bet I won't pass it by.

23. Billy

My first military leave home was exciting on many levels, not the least of which being that I would be seen in uniform. On another, my sister had arranged a date for me with someone whom I had long admired, but who had never been available when I was.

My family had moved since I entered the military, and the route to my date's home was long and unfamiliar. When I entered a village called Welchville I knew I was a good half hour behind, and it was here that a back tire blew.

I pulled over into the well-illuminated parking lot, got out and opened the trunk. The snow was falling lightly and the spare tire looked very dirty. I was concerned about my spit-shine and dress uniform.

Across the street was a coin operated laundromat and out of it came a tiny, white-haired, old couple.

They loaded their laundry into a car, their eyes on me all the while. Then they drove across the street and parked behind me with their headlights on my work. They got out and offered a flashlight, which was quite unnecessary.

It was November and bitterly cold, there was really no way they could be of help. I felt pressed by the thought that I should already be at the door of my date. I viewed the old couple as a bother and suggested that they should get out of the cold.

However, they wanted to be there.

They stood behind me in silence, holding the flashlight, huddled together with their arms around one another as the snow built up on their heads and shoulders. I felt the atmosphere getting heavy and I wished they would leave.

"Doesn't he remind you of Billy?" the woman asked her husband.

The old man was silent for a moment, then he sighed deeply and said, "Yes, very much."

Now I'm going to have to hear all about Billy, I thought, looking at my watch. I wished they would be sensible and get out of the cold.

Then the two of them told the story of Billy who had gone away in a uniform like mine many years before, but who had never returned. It was more like a double soliloquy than a conversation, as for a short time the old couple re-experienced Billy through me.

I reluctantly relaxed to the gentle emotion of their sad story and listened as I did the work at hand. Twenty minutes later the tire was changed. By that time the old couple had let out enough grief that I didn't feel rude about leaving them. I thanked them for their help and we parted ways.

I was an hour late for my date and her mom put a midnight curfew on us. We dashed off to a dance an hour away, and I don't remember much else about the date.

In the time since then, it is my encounter with the old couple which stands out as the main event of that evening.

You see, years later I chanced on a short story by some Russian author. The chief character was a horse-drawn taxi man.

The setting was a cold, snowy evening in a Russian city center. He had several patrons during the evening and to each he tried to talk about his son, but everyone was too tied up in their own affairs to listen.

Finally, late in the evening, he returned home, put his horse in the stall and went into the house to sit by the fire. After some time of staring into the fire, he lit a lamp and returned to the barn.

He sat down in front of the horse and addressed him, in these words: "You know, I had a son, a wonderful son … twelve years old … but he died a while back, and I need to talk about it."

How thankful I have been since reading this story that I took the time to listen about Billy.

were much impressed when they saw my sideboard. They had to get confirmation from the auctioneer before they would believe that I had actually bought the thing. Grandmother was delighted and that clunky sideboard remained useful on her piazza until old age required her to give up the farm.

So I had learned firsthand what an auction was, and I had also learned that one should never bid on something one does not truly want.

27. First Kill, Last Kill

My first deer kill was at age twelve. It was as much a case of self-defense as anything else. Dad jumped a barren doe from a pine thicket. He fired. She came bursting out of the pines headed straight for me. With fear I raised my gun, she turned oblique left, I fired.

The bullet pierced her heart but she was still alive, and while she was, she ran her life out, passing through the opening where I stood and disappearing into a stand of fir.

There was a dusting of snow on the dead oak leaves making her bloody trail easy to follow.

I encountered a small Christmas tree over which she had passed. It was red with blood. I found her a few yards into the firs. She lifted her head for a brief moment, looked at me with soft eyes, and died.

From the time the bullet struck until she was dead was no more than two minutes. It was about as quick and clean as such a death could be. I felt good about the kill and competent. It was for me a rite of passage.

The next year I experienced quite another sort of kill. In Maine we were hunting the Otisfield Gore section called the Flat Iron. A skipper dashed across a neglected field filled with small pine. Several hunters in our party fired at her. She went down.

In a moment she was up but running very low. I could

catch glimpses of her back now and them. Her movements were not graceful they were the disjointed hobble of a wounded animal, the body language of pain. I saw her cast over a distant wall.

I knew my uncle would be waiting. She crossed the road, Uncle took aim, his rifle cracked, sending a piece of lead through her neck. The doe dropped dead in a ditch.

I approached the kill to find three of our party looking down at her in silence. The first volley had blown off eight to twelve inches from each of three legs. The one remaining leg had been too long to be useful. She had been running on broken-off bones.

The marks crossing the dirt road were not the marks of hooves, but of bones—bone prints. The centers of her bony stumps were filled with mud and moss in place of marrow.

The image of that delicate deer, her drive for life drained out of her, her mutilated gentleness dead in a ditch, took from me forever the thrill of the chase. And out of reverence I have put up my guns from hunting. Out of reverence I have locked them away.

28. Lambchop

Amongst my most vivid childhood memories are those centered around my 4-H lamb projects. Early in the spring we'd all take a family outing to a local sheep farm to select a young ram for fattening. Lambs were required to be about six weeks old and castrated. I never understood just why castration was a requirement, but then there are many things a lad of ten doesn't understand.

Project participants were required to raise the young ram, keeping detailed records on feed and other expenses. Then in September we all amassed at the county fair with our now large, fat lambs to be competitively judged and awarded ribbons.

The following day we gathered before the fair grandstands, where one by one we led our project animal to a make-shift stage and watched as bids were mysteriously received from the mass of faces assembled for the purpose.

"Sold, 23 cents a pound," shouted the auctioneer, slamming down his gavel. The project was over. We 4-H'ers departed for home, leaving our animal behind, and in a few weeks a check arrived in the mail.

There was another mystery here for me: who bought these animals anyway, and what did they do with them? No one in my family ate lamb, nor did I know of anyone who did.

Years later the military assigned me to Pakistan, a land

where the favored animal protein source is lamb. While there, I learned to greatly savor the special flavor of lamb; it became my preferred meat.

After my military service I returned to the University of Maine and found that my now favored meat, lamb, demanded 2 1/2 times as much money per pound as did the cheaper cuts of beef. I was on a chuck steak budget, and lamb became a rare luxury in my diet.

After graduation my wife and I, during late winter, moved to a farm a few miles north of Portland, Maine. We were eager for spring to arrive so that we could plant a vegetable garden, purchase a flock of hens, and a spring lamb to raise for the freezer.

In April, much to the delight of my wife and daughter, I brought home an eight weeks old, castrated, little ram. So that there would be no doubt as to the end of this wee creature, I named him *Lambchop*.

Lambchop became a favorite of the whole extended family and the neighborhood. Children loved to visit our backyard to romp and play with this bouncing bundle of wool.

As summer waned, Lambchop became a good-sized animal. Folks knew that his departure was imminent, and some were critical of the idea.

One Saturday I announced that the time had come for Lambchop's transformation. Word spread quickly. My neighbor, Ghunnar, arrived with a message from his teenage daughter: "If you take Lambchop to the slaughter house I'll never speak to you again."

In my memory she hadn't spoken to me during the nine months I'd been her neighbor, so I counted it as no great loss. I took Lambchop away and brought back neat little packages for our freezer.

I was disappointed to discover that my wife and daughter would not share this delicacy with me. It took all year for me to eat it by myself.

Spring arrived again, but I was very busy and by the time I went shopping for spring lambs they were in short supply. Finally I located one, but the farmer hadn't had time to castrate the animal. He warned me that I must attend to it right away.

I took the young ram home, got occupied with various other projects, and suddenly remembered that the animal hadn't been altered. By this time, however, he was 12 weeks old, making the operation a more serious and costly undertaking. I decided to let the young ram remain as he was.

This lamb was as popular as Lambchop had been; however, when I named Lambchop I had associated the name with a dinner plate, not a fuzzy little TV puppet. I decided to stay away from cuddly-cutsie names this time. Butch, this ram became.

Friends and family joyfully frolicked with Butch throughout the summer, but as September leaves fell Butch began to feel his oats.

He first manifested this endocrinological change by butting things. Once he butted our Labrador Retriever so severely in the ribs that the yelping dog ran off into the woods and didn't return for three days.

Then Butch took to butting people. My wife now referred to the former cuddly lamb as that animal! Children viewed our backyard as a dangerous place to be.

Then one evening I returned home from work to be confronted by an irate wife. "When are you going to take that disgusting animal to the slaughterhouse?" she demanded. "He's been mounting a beach ball all day. Now that's a fine thing for the children to watch. That animal is a menace!"

I knew the time had come to take Butch on a one-way journey, and I also knew that this time I'd be able to do so without raising the ire of family and friends.

The following Saturday I took Butch for a ride, brought him back in smaller portions, and stored him in our freezer.

I was pleased to find that this year my wife and daughter were willing to sample some of the meat.

We developed a Sunday ritual of lamb for dinner and this delicious protein source became a favorite for the whole family.

I had also learned a valuable lesson here: never again did I find myself with enemies at slaughter time, because never again did I have my freezer lambs altered.

29. *Linwood*

My family in Maine recently sent me a local telephone book of Oxford County so that when I wanted to call someone back home I wouldn't have to pay those exorbitant information charges.

One evening I amused myself by looking through this phone book for the names of people from my childhood to see who was still around the old stomping grounds.

Much to my surprise I spotted the name of Linwood Bishop who had been in my first grade class. His father had been a single-parent woodsman who had done his best to raise three children. I remember the Bishop kids as being a very bright, but rough and tumble bunch.

I recalled a December day some 40 years ago when a thick moist snow fell. We kids rushed out at recess to make our first snowmen of the season. Linwood was fast and soon asked me to evaluate his creation. What I saw so shocked my Sunday-school-sensibilities that I spontaneously kicked it over.

With a stick and some mud, Linwood had produced the most graphically pornographic image that I, thus far in my sheltered existence, had seen or imagined.

Linwood jumped me and down we went in a hitting, kicking flurry. Then I felt a sharp pain in my left arm—Linwood had bitten me. I screamed out in pain.

Suddenly we were jerked to our feet by the collars

and were marched into the warm schoolhouse to face our teacher, Olgar Bean, who demanded in measured cadence an explanation for our behavior.

"Linwood bit me," I protested.

"Is that true, Linwood?" she asked.

"Yes." replied Linwood defiantly. "He kicked over my snowman."

"Did you kick over Linwood's snowman, Henry?" she asked.

I was beginning to feel cornered. Silently I nodded my head in anticipation of the dreaded question.

"Why did you do that?" she pressed.

I couldn't tell her, I couldn't have told anyone about that snowman.

"Well, Henry," said Mrs. Bean in her typically impeccable logic, "if you don't want people to bite you, you shouldn't kick over their snowmen."

Case closed.

Linwood moved away shortly after that and his name in the phone book was the first I'd heard of him since.

I wrote a letter to this Linwood in the phone book and asked if he were the same Linwood Bishop who had bitten me so long ago, because we needed to discuss it.

He wrote back saying, yes, indeed he had been my classmate in 1950, and though he had no memory of biting me, he apologized for doing so. I called him up and we had a talk about the good old days and shared some things about our lives since.

Now I can do closure on that long past event, and I've renewed a friendship.

30. Lost Mystique

My elementary school was located on a three-acre lot with deep woods on two sides. A stonewall surrounded the property and marked the limit of our territory. Crossing that wall was strictly forbidden. There were, however, two big attractions just over the wall in those woods, and the absence of windows on the backside of the schoolhouse permitted an occasional breaking of the rule.

One attraction was a pile of huge boulders. They were piled up in such a way that on the backside there was an opening, a cave to the imaginations of us school children. To sneak over the wall for a peek into that blackness was one of the most adventurous activities that the children of Maine's Otisfield Central School enjoyed.

We didn't know exactly what lived in there, but we certainly had some theories. Many claimed that it was a bear den, but those of us who were wiser knew that it was too small for a bear, and besides, bears are not very exotic in Maine. Most of us found it more exciting to believe that a ferocious wolverine had migrated into our area and had taken up residence under our pile of rocks.

No one had ever seen the beast that lived in the black hole, but many could testify to evidence that something lived there.

Every now and then we seemed to need re-convincing.

One team would be dispatched to keep a lookout for teachers, another would venture to the wall, and eventually one of the braver kids would cross over to have a look.

The adventurer inevitably returned with wild-eyed testimony to having heard a growl, scratching, rustling, or perhaps they saw evidence of tracks, big ones. We, the exploring party would then break into a run for the safety of the swing sets, fearful that the monster was at our heels, and then we would enjoy the complete attention of our peers as we shared the evidence that some fearful creature lived so very close to us.

How exciting it was, every fall when school started, to re-verify the existence of the creature under the rocks.

And there was another great treat waiting for us on our return to school—just over the wall, on the other side of the school grounds, there was a marvelous wild apple tree. I call it an apple tree but in truth it was more like a grove of apple bushes. About thirty trees grew in a clump of long willowy branches, a spray of apples. The apples produced were golf ball in size, yellow and scabby. The fruit became ripe just as school started, and they had a taste that only a little boy could appreciate. They were very juicy and wonderfully tart and surprisingly worm free. Boys' apples, I call them, because none of the girls, as I remember, would eat them.

Obtaining these apples was a bit more risky that peeking into the den, in terms of getting caught, that is, because one of the classroom windows faced the same side of the school grounds.

The safest approach was to cross the stone wall at the back corner of the property and then scoot along, crouched behind the back side of the wall until we were in the vicinity of the apple bushes. We then waited for a signal that the teacher was at her desk, from where she could not see the point of action. Then two or three of us brave lads would dart into the apple bushes, grab a trunk, give a few shakes, and hurriedly pick up the bounty.

We then dashed to the cover of the wall and returned to the cross-over point, the pockets of our jackets and trousers bulging with a tasty treat to be shared by all.

I am sad to report that these two wonders of the Otisfield Central School are no more. You see, from the point of view of the adults, improvements were made at the school.

There was a fine old man who lived nearby. He and his wife had had no children, and I suspect he was thinking about this when he decided to do a special favor for the local children. He owned the property on the other side of the wall from the schoolyard, and he thought he could please the children by enlarging their playground.

One summer he brought in a crew of loggers and all the trees were cut down, including the sacred apple bushes. Huge bulldozers pushed up the tree roots and with them pushed the wolverine den to the far end of his property, where all was buried.

At the same time, in front of the school, the kind old man implanted tether ball poles and constructed wooden jungle gyms. You see he was a carpenter, and he gave to the children what he could do best. His intentions were entirely kind. The teachers, the school board, and the PTA were all wildly appreciative when the old man turned over the deed of his property to the school.

None of them knew anything about the animal den or apple bushes.

We children found it difficult to say thank you. We had returned to school in the fall to find that the wonderfully scary den of creatures and the deliciously sour apple bushes had been replaced by a flat field devoid of mystery and wonder. We felt loss and sadness and found little compensation in the playground the old man had so carefully constructed for us.

Something very special had been lost. The atmosphere, the source of excitement, the mystique of the Otisfield Central School had been greatly diminished. It had been replaced by a thoroughly unexciting four-acre field.

Shortly after this, the town joined a school district and the kindergarten through eighth grade students at the school were replaced by solely second graders. The new teachers of these young children didn't want them wandering off into the field, so they put restrictions on how far they could stray from the school house. The imaginary line they drew was just about where the old stone wall used to run, just short of the spot where the wolverines once gathered.

No apple has ever tasted so wonderful to my tongue as did those which once fell from the apple bushes, whose roots are now buried under the distant mound. For me the ghost of those bushes and the den still stand in that field, its blandness mocking their past glory.

31. Marion

I've enjoyed spending the summer in my hometown where I've encountered people from my distant past as well as the memories associated with them—Marion, for example.

I remember Marion as a spunky young girl with beautiful eyes and rich hair. She was a bit uppity, but at the same time, somewhat of a flirt. These traits together attracted my interest considerably.

I must confess that Marion was often the victim of my somewhat addlepated monkeyshines.

On one occasion in the school lunch line, I had a packet of alum in my pocket—Grandmother's remedy for mouth cankers. I opened one of the bread and butter sandwiches stacked there for our taking and sprinkled on a bit of the potently bitter powder. I took my lunch to one of the long tables and waited for a reaction.

"Rancid butter!" choked Marion, spitting out the bread onto the table. I kept my head low throughout the commotion that followed and silently experienced regret that pretty Marion was the victim of my secret prank.

Another bit of tomfoolery took place in that lunch room when once, after taking my place at the lunch table, I noticed that just over the straws that stuck out of the milk bottles placed there for our convenience, hung a little spider. I steered its filament so that it landed on one of the straws. The frightened spider crawled down into the straw, a circumstance I accepted with great delight. Who would ingest a spider with their lunch? I wondered.

Soon Marion sat down across from me.

She immediately snatched up the loaded milk and drew on the straw. I watched the dark spot in the straw ride the column of milk up into her mouth. She was unaware that she had consumed a spider, but I sat there feeling guilty that Marion, whom I admired, had once again fallen victim to my trickery.

In those days girls wore copious petticoats, and because the stairs leading down to the lunchroom were very steep, petticoats trailed three steps behind the young ladies. One day while Marion was descending the stairs in front of me, I was overcome by the temptation to step on her fancy train.

The petticoat stopped immediately, but by the time Marion stopped about a yard of petticoat material hung from her hem. She looked up at me with her beautiful Italian eyes and blushed.

Then she calmly tore off the ripped material and casually dropped it into the trash can at the bottom of the stair.

Her cool reaction to my meanness made me feel guilty again, and I resolved not to play any more tricks on Marion.

But alas!

Marion went to a private school the next year and I didn't see her until the following summer. She was at the public beach and when I approached her, Marion screamed and ran off down the beach, accusing me of hiding a frog in my closed hand. I was innocent of the charge, but she gave me an idea.

I would find a frog and conceal it in my mouth.

I would walk up to Marion with open hands and smile, and then I would spit the frog on her.

I located a green frog with brown spots, a leopard frog. But as soon as I popped it into my mouth, my mouth was filled with fire. I spit out the frog and ran to the lake for relief, frantically swishing gallons of water through my burning mouth.

Years later in college zoology class I learned that the leopard frog is unique amongst frogs. The brown spots are not just spots, they are glands which secrete acid to discourage predators from swallowing them, and little boys from hiding them in their mouths.

That was the last time I ever tried to pull a prank on Marion. Perhaps, I had come to realize that I was as much a victim of my shenanigans as was she.

32. Mothers Have to Think Quickly Sometimes

Mothers have to think quickly sometimes. I recall one Christmas when I awoke in the wee hours of the morning, wondering if Santa Claus had yet come. I drew back the curtain to look outside for evidence of tell-tale tracks in the snow. The sky was as clear and as star bright as only it can be in a northern winter night. I saw no evidence of Santa in the snow, but high in the western sky in the middle of an exceptionally bright gathering of stars, I swear I saw a distant sleigh and a string of reindeer.

Here comes Santa Claus, I thought and was filled with panic. Everyone knows that Santa passes by those houses in which the children are not sleeping.

I rushed to my parents' room and startled my mother awake. "I saw Santa coming," I blurted. "I'm awake; he won't stop here."

Mother pulled me into her bed and whispered into my ear that she would tell me a secret if I would calm down.

Then she explained that what I had seen was not Santa coming, but Santa going. "He was just here," she said. "I heard him putting your gifts under the tree. Now you go back to bed and get some more sleep."

I did so without delay or anxiety.

Mothers have to think quickly sometimes.

And then there was Scotty's Easter. Scotty is my kid

brother. One Easter he awoke and eagerly searched the house for the expected chocolate bunny and jelly beans, but search as he might he could find none.

Then Mother was awakened by sounds for which mothers' ears are especially designed—a child's whimper. She sat up to see Scotty, a big pout on his face and tears streaming down his cheeks.

"What's wrong?" she demanded.

Scotty found it difficult to speak, but finally, in a burst of tears, he was able to explain, "The Easter Bunny didn't leave me anything!"

Mother's heart sank as she remembered what she had forgotten to do the evening before.

She rushed from bed and embracing Scotty assured him that he must be mistaken. "You know," she said, "you're a big boy now, and the Easter Bunny probably hid his gifts very well."

Scotty's fears were calmed considerably with this information. Mother took his hand and together they looked in all the hard places where a bunny might hide things. They looked under the sofa, in closets, drawers and cupboards.

Finally, Mother opened the high cupboards over the kitchen sink. There she found a big brown paper bag and setting in on the floor in front of Scotty, she said, "Look at that, the Easter Bunny pulled a trick on you."

Scotty cackled with delight and relief as he recognized chocolate ears sticking out of the bag, then he saw marshmallow chicks, brightly colored eggs, and countless jelly beans. The Easter Bunny had not let him down.

Many days passed before Scotty tired of telling everyone

he met, relative, friend, neighbor, or complete stranger all about the big trick the Easter Bunny had pulled on him. That rascal's trickery had made Scotty feel uniquely important and grown up.

Yes, mothers have to think quickly sometimes, and I think that my mother was particularly good at it.

33. A Less than Traditional Wedding

My wife and I were married in 1986. It was our second marriages and differed considerably from our traditional first marriages.

Now, I have never cared much for traditional weddings. I find them stuffy and formal—women's affairs. Reception lines in particular I abhor. Usually most of the guests know only the bride or groom, yet all these people must line up feeling ridiculous in their monkey suits. Hey, this wasn't their idea.

I tolerated such formalities the first time. I viewed it as an ordeal I had to go through. My input in the affair was zilch.

The second time, however, was a mutually arranged celebration, and a celebration it should have been. Both of us had been 7 to 8 years between marriages, and this one was a highly conscious decision. I resolved to making it the biggest shebang I'd ever organized.

Our less than traditional vows took place by my garden pond with thousands of goldfish as witnesses:

"Are you willing to love Shirley into her unique fullness and to take the risk and the vulnerability of love again, and again, and again?"

I was, so we eagerly took off as husband and wife to the bash … er, ah … reception, which the good Reverend excused himself from.

Wedding receptions can require exhausting arrangements, months of planning, but we threw ours together in two weeks.

Everyone wants to be married on a Saturday in June so halls, bands, and caterers are all at a premium at that time. We chose Sunday. We found little competition for wedding services on Sunday.

We wanted to incorporate into our celebration those aspects of our lives that we took particular pleasure in.

Athletics for example. The reception was at an indoor tennis club, complete with racket and walley-ball courts where those so inclined could celebrate our union with activity on the courts.

For those who wanted activity on the dance floor, we hired a rock and roll band. There is not much demand for a rock and roll band on Sunday afternoons. Our first choice was available.

We also hired a bagpipe band to share the love of our Scottish heritage. We wore traditional Scottish attire, imported heather from the Highlands, managed to get nearly everyone involved in Scottish country dances, and drank a toast with single malt scotch.

My Mid East experience convinced me that yet another type of dancing was required: belly dancing. That's right, no wedding I attended in that exotic part of the world was complete without a belly dancer.

Mother was a bit shocked at this idea, but the actual event was a tame shadow of what she must have originally imagined.

Now, what do you do for catering on short notice?

We called Dominos and ordered 20 big pizzas for one

o'clock and 10 more for two.

Now before you get the wrong idea about why we were in such a hurry to get married, I must interject here, that we had both been accepted by the Peace Corps. When we notified Washington that we had decided to marry and to go in as a couple, they insisted that we be married for at least six months before they would assign us as a husband and wife team. So the sooner we married, the sooner we'd get an assignment.

Anyway, I had a wonderful time at my second wedding, and I'm sure the guests did also, not to mention the non-guests—we found a number of mysterious faces in the photographs of our reception.

When the celebration ended the leader of the rock and roll band told me that ours had been the most organized wedding he'd ever done.

The truth was, it wasn't organized much at all, it was a simple down-to-earth spontaneous and joyful celebration, as has been our life together ever since.

34. Respectful Progress: The Old Oxford County Fair Exhibition Hall

My first experience of the Oxford County Fair as an unescorted (by parents) youth was in 1959. I entered Paris High School from Otisfield that year. To escape the issue of kids skipping school to attend the fair, both Norway and Paris High Schools suspended classes for a day, permitting students to attend the fair, which was within walking distance from both schools.

A number of experiences that day stand out in my mind. I kissed a girl in the darkness of the Haunted House, and for the first time in my life I invited a girl to join me in a Ferris wheel ride. Her nickname was Bones!

Then, as I was throwing softballs at cast-iron milk bottles, a former classmate from Otisfield, then at Norway High School, stopped with his girlfriend and her friend. The body language of the pretty friend made it clear that she was not interested in my joining them as her escort. Not until, that is, I threw a ball and knocked over all three bottles. The attendant handed me a stuffed animal—a great gift for my baby brother, I thought.

"Oh, thank you," exclaimed the pretty girl as she snatched my prize and hugged it appreciatively. At that I lost all interest in making their threesome a foursome. We parted ways, and off went my brother's gift in the arms of a girl whose beauty was forever tarnished for me.

But my most vivid memories of the old Oxford County Fair are centered on the Exhibition Hall. I was a 4H-er and displayed with pride my creations and vegetables there, back when cookies and pickles were judged by taste rather than looks.

My mother didn't serve us fried food much so I was never attracted to the most popular fair foods. My favorite place to eat was just inside the large doors on the Exhibition Hall's southern entrance. There, local ladies served their homemade cooking. On that day I sallied up to the counter and ordered my favorite, blueberry pie. Mrs. Levine, mother of my twin classmates, Michael and Karen, served me that pie and a glass of milk. I was not disappointed.

What was so intriguing about wandering about a great hall, looking at displays of fruit, vegetables, handiwork, and collections all judged with appropriate ribbons? I'm not sure. But I enjoy it still, judging the judgments of the judges and looking for entries from people I know.

The great hall was an appealing place in itself. Its high ceilings, framed timber structure exposed, its wonderfully creaky floors and stairway all contributed to something more real and genuine than the tents, plastic, plaster of Paris, and noise of the fair outside its walls. It was the same building year after year, and its essence sank into one's psyche. It became an important symbol of our home stomping grounds, our roots.

I graduated and wandered the world, returning recently to find a regional high school on the fairgrounds site but the old Exhibition Hall still standing. It had been gutted and its clapboards ripped off, but its old form was most recognizable. And now it's back together, serving as a community college, a university, and a career center, all respectfully placed inside this symbol of the past for so many people. Much of its timber structure can still be seen, and some of its former creaky floorboards form a wall. The past has been transformed into the present, and its old essence can still be felt within its walls.

We have here an example of respectful progress. If you haven't been in to check it out I suggest you do so soon.

Western Maine Community College requested this piece from the author.

35. The Skeleton

My great-grandfather read the Bible constantly. He professed to have it memorized from front page to back. This might have been the truth, but I'm not sure because the family enjoyed bragging about Great-granddad's religious feat, and there was no one around who could or cared to test his claim.

For over a decade at the end of his life Great-granddad was bedridden. When my family went to visit my eldest aunt with whom he lived, I was obligated to pay my respects at his bedside. The event was not pleasing to a child. His room was dark and the odors there disturbing. His touch was cold and clammy, his wizened face ashen. He was a dying man.

He always mentioned that he had the Bible memorized and wanted to know if I read the Bible. I would invariably get a lecture on the importance of daily Bible reading. I found the experience intimidating and yearned to escape outside to where my cousins awaited for fun and games.

I really never knew the man he was. I remember him only as a near corpse with bed sores and strange odors.

Great-granddad died while I was in the military. A decade passed before I visited my aunt's farm again. I was then wiser to the ways of the world, and I said to my aunt, "I have often wondered what Great-granddad did as a young man that caused him to be so fanatically religious in his old age. What do you suppose he was trying to make up for?"

"You know," said Aunty, surprised by my question, "Grandpa said the most astonishing thing from his death bed. I've never mentioned this to anyone, and I'd nearly forgotten about it. On the last day of his life he told me that he had to confess a great sin."

"Now Grandpa," Aunty told him, "a man like you has no sin to confess."

But Great-grandpa protested that he did, that he had committed a profound sin. He then told of his marriage in the early 1880s to a woman eager to have children. But when the turn of the century arrived, his wife still had produced no offspring. This became a great torment to her. She became depressed and spent long hours weeping. She lost her appetite and became dangerously thin.

During that period Great-grandfather went out into the mountains to hunt moose. From a mountain side he saw smoke arising from a place that had been virgin forest when he last was there. He speculated that an Abenaki Indian hunting party was cooking a moose, so he made his way to the spot to partake.

What he found was a clearing with a new log cabin. Smoke curled from a new stone chimney. The front door was open. He knocked and called out, but no one answered. He shouted about the clearing, but got no response. Then he went into the cabin, and in a crib he found sleeping a newborn babe. He spontaneously snatched up the sleeping child and ran off into the woods.

He returned home with the child and presented it to his grieving wife. She was overwhelmed with joy. Months later they traveled to town and registered "the birth of their new child."

"That child," explained Great-grandfather with his last words, "was your mother, and I stole her."

After relating our great patriarch's deathbed tale, my aunt turned to me and said with emotional disbelief, "Now why do you suppose he ever said anything so crazy as that?"

36. *The Spirit of Places*

Children know about the spirit of places: the basements of abandoned houses, the attic in grandma's house, cemeteries, and church sanctuaries. Each has its unique spirit.

As we grow older, we usually convince ourselves that the various feelings that places give us are nothing but projections, a fear of the unknown. A basement becomes just a basement.

A number of life experiences have rekindled my belief in the spirit of places. I once had a job which required me to go into many people's homes. Some homes gave me happy feelings, some peaceful; but two, I remember, filled me with unaccountable dread.

Later I learned that terrible abuse and perverted activities had taken place in those houses. Their spirit had been tainted by it.

In 1985 I toured Scotland with my children. We visited some two-dozen castles, but Dunnottar near Stonehaven was the most memorable in terms of its spirit.

Dunnottar Castle is a large complex built over the centuries on a sea cliff. We entered all of the many buildings—except one, the original tower, or dungeon. As we looked into it we felt dread and foreboding. We could not bring ourselves to enter in.

Later I read of the horrible crimes against humanity that had taken place there: torture, maiming and murder. The souls that languished and died there so long ago have left their imprint on the soul of the place.

In 1990 I took my son to Chattanooga. From nearby we rode the Incline Railway to the top of the mountain and wandered up the hill to where we found a park.

Now forgive me, I grew up in Maine, far from the scenes of the Civil War. The name of Lookout Mountain meant nothing to me before that day.

As I approached the gate I was overcome by feelings of sadness. When I wandered about the park, reading the plaques which described what had taken place there, I was not surprised to discover that the site had a tragic history.

Thousands of men, neither from meanness nor from malice, but from belief and loyalty, had slaughtered one another there, a living force of solemn sadness still pervades the place.

Once while hiking deep in the woods of central Maine, I came across a well-worn path. Out of curiosity, I followed it.

I came upon a circular clearing, about 1/4 acre, where all the brush and trees, save one, had been removed.

The one tree, an oak, stood in the center of the opening.

Many years previously, the tree had been partially uprooted by a storm. At about nine feet it had been nearly snapped in two and bent over to the earth. The break had healed and the top had grown up again. It was a zigzag tree.

Under the elbow formed by the break, a bubbling shallow spring had been dug out and rocked up. There was a goldfish in it. Wood flowers grew in profusion out of the moss covered earth, and a log served as a bench.

Someone loved that isolated spot in the deep woods and

spent a great deal of time there in communion with nature's splendor. Human loving care of this place had given it a rich, Godly atmosphere, and I felt at peace in its aura.

Yes, it is people, and what they do in a space that gives the place its spirit.

37. Maine Sets Traps for Man Wanted by FBI

This was the headline from the *Maine Sunday Telegram* newspaper, February 23, 1959, and I, a seventh-grade detective, read it with the interest of a professional snoop. I was in the Otisfield General Store at the time with fellow detectives Dana and Craig Grover and Herb Wiley. We were on our way to slide Cobb's Hill. But first, we read about "the flying bank robber," Frank Sprenz, who was hiding out in Maine.

As we walked along Cobb's Hill road, our imaginations were consumed by visions of capture and reward. We speculated that the wanted man might be held-up in one of the summer camps off that very road. From the base of the hill we noticed a car stuck near the top.

As we approached, we observed two men. One was a wiry character with a rodent-like face. He was behind the car in shirt sleeves shoveling angrily at the snow. Beside the car stood a hulking man in a black trench coat and broad brimmed hat. What struck us country boys as strange about this city slicker was that he wore sunglasses in the middle of the winter—suspicious!

I was leery of these strangers and intended to walk past in silence, but Herbert, who was more comfortable with adults than the rest of us, made an attempt to be friendly. "What's the matter, stuck?" he asked.

The ratty little man threw down his shovel, jumped out of the ditch, called us "wise guys," and then proceeded to abuse us with language that would have gotten our mouths washed out with soap had our parents heard us using it—we were frightened dumb.

We continued up the hill with excited minds until we were over the top. Then we turned to one another and in one voice exclaimed: "Do you know who THAT was?"

We all knew, knew without a doubt, that we had encountered the infamous "flying bank robber," Frank Sprenz and his accomplice, Rat Face. We managed to get back past these criminals on our Super Flyer Sleds and ran to the local store to tell our story.

We were pleased by the interest with which our tale was received. "What kind of car was it?" we were asked. "A red and white 55 Chevy station wagon," said Herb who knew cars. We didn't know that the state police had just called the store to warn that Sprenz and a companion had been spotted in the Norway Liquor Store ten miles away, and that they were driving the very sort of automobile we had described.

When we departed for Herb's house to draw sunglasses on the photo in the newspaper, the proprietor of the store, Mrs. Butler, called the state police and the local deputy.

Now, a lot of folks dropped in at the Otisfield General Store on Sunday afternoons, and each one who did on this day heard the story about who was stuck on Cobb's Hill, and, of course, they stayed around to see the action. Soon there was a great crowd in the place.

You can imagine the chill or terror that swept through this throng when the jingling door admitted a stranger in black, the Devil himself! He found the little country store full of silent staring people. Unseen, the local deputy reached under his coat and put his hand on a revolver—both the deputy and his revolver were loaded.

The stranger wanted change for the phone, got it, and passed through the stony silence to the phone booth outside.

When the door closed behind him, the good folk of Otisfield exchanged looks, confirming to one another that a fiend had been in their presence. My father, who was one of that silent assembly, crossed the road to the Wiley farm to take me home before the shooting started.

You see, the state police were converging on Otisfield. They had called again to warn that Sprenz was armed and dangerous and that no one should approach him.

Herb and I saw my father coming and dashed out of the farm house with the newspaper photo of the crook; the added sunglasses, we believed, proved our case.

Dad didn't need convincing. He frantically gestured SILENCE, and jerked his thumb toward the dark figure who was trudging back toward Cobb's Hill. Probably he found the chill there preferable to the one in the little country store.

The troopers finally arrived—a veritable army, I'm told.

The two lonely men on Cobb's Hill were quite astonished to hear themselves addressed through a bullhorn. They looked around to see rifles pointing up the hill at them and rifles pointing down the hill at them.

They surrendered without resistance. They were yanked around and frisked. They presented the identifications of completely harmless men which, when checked out, proved valid—a near blind man and his lost chauffeur.

Perhaps my friends and I did not receive any material

reward from all of this, but we did experience considerable excitement and notoriety. The crowd at the store enjoyed chuckling amongst themselves about the foolishness of boys. And as for the two out-of-staters ... well, I'll bet you they thought twice before they spoke so rudely to country boys again.

38. Confession 50 Years Waiting

A brook ran between my childhood home and Grandfather's farm on Rayville Road in Otisfield, Maine. Over the centuries its erosive power had created a gentle valley, permitting me in the winter to slide its grade from my yard to Grandfather's.

An ancient fieldstone bridge provided passage over the water's way. In the cool atmosphere under that bridge, when summer had reduced spring's surging stream to a babbling brook, I found a primary place for my summertime fantasies. It became a cave to explore, a pirate's den, its walls offering hiding places for the treasure gleaned from the rejected costume jewelry of grandmothers and aunts.

Road crews appeared one summer with earth moving equipment. The authorities had decided that the road dipped too much at the bridge. They trucked off the huge slabs of granite, cut and laid by early settlers, and laid down in their places a long four-foot steel culvert. They brought in loads of gravel and built the road up six feet, filling in the offensive dip, leveling the grade that had permitted wintertime sliding.

The long black tube had little charm. Its tar covered corrugations didn't hint of caverns or dens. No self-respecting pirate or Injun Joe would hold up in such a bland environment of industrial uniformity. Townspeople gathered to admire the improved road as I pined for the loss of wintertime sliding and summertime adventures with pirates and thieves.

It was a dry summer that year. The stream shrank to a tiny trickle. My ducks had no place to swim, and I feared that they would abandon me, as they had in the past, for an upstream beaver swamp. Then I saw the solution to my problem. A short 2X8-inch plank, carried there by the water's springtime force, lay in the stream bed. I placed it across the upstream end of the culvert. Within a few days my ducks were joyfully frolicking in eight inches of water.

I retrieved additional planks from Grandfather's shed to block off more of the culvert. Eventually I'd created a three-foot dam across the culvert's opening, backing up a veritable pond of water. Countless trout took refuge in it from the dying brook. Deer, mud hens, wild ducks and muskrat were likewise attracted. My creation was a great source of summertime pleasure for me, and in the fall it froze over, offering me a skating rink.

Thinking back, I've often wondered why the adults who traveled that road every day never questioned the sudden appearance of a pond—so much water in a dry summer? Things might have gone better had an adult investigated the situation, because an adult would have foreseen a danger that escaped me.

Spring arrived, and with it the warm rains melted the deep snow, sending torrents of water into the stream. The meager space I'd left at the top of the culvert's mouth was an insufficient conduit for all of that water. The stream rose over the culvert and then rose another six feet over the road, creating a huge lake which reached our house.

Deep in the night a great roar awakened my family from sleep. First we discovered our basement flooded. Then Father went outside to investigate. He returned to report that the new road had washed out.

I did not connect this circumstance with any of my actions.

The culvert had stayed in place, and unfortunately, so had the incriminating planks. When I returned from school the next day, Mr. Peaco, the road commissioner, and my father confronted me with the evidence. I fixed my eyes on their feet and said nothing.

Ten-thousand dollars damage, they told me, more money than my father earned in two years. I was so distressed that I nearly vomited on the road commissioner's boots.

Mr. Peaco, his big white belly protruding between his shirt and pants, pulled away the remaining planks from the culvert's end and let the stream carry them from view.

"I won't mention this to the town fathers," he told my relieved parents.

Climbing into his dump truck, he looked at me sternly, saying,

"I hope you've learned a lesson from this, boy."

I had.

When the road had been repaired and the next summer began to take a toll on my precious little stream, I put only one plank across the culvert's end. It blocked enough water to keep the trout and my ducks happy.

In the fall I removed it.

39. The House of the Rising Sun

A jazz rendition of "The House of the Rising Sun" is playing on the radio and I am reminded how certain pieces of music can bring back past events of life with poignant vividness.

Sometimes the association isn't pleasant. During the demise of my first marriage, for example, I found the tone of the Shostakovich 5th Symphony to fit my mood precisely. For me the gloom and pain of those days are bound up in this music.

It seems that when music is associated with an intense or new experience the two fuse inextricably.

I remember a moment of intimacy when a lady friend turned on the radio and a velvety voice was singing, "Are you Ready?" I always enjoy hearing that song.

Christophen's "Sailing" takes me back to a particular dance floor where I held a lovely lady as we danced in a trance, wishing that waltz would never end. That was a rich moment.

But then there is the peculiar association I just can't get out of my mind when I hear "The House of the Rising Sun." It takes me back to my first military assignment abroad—Chicksands AFB, near Bedford, England. My roommate there offered to show me around town, and we ended up at the most popular pub in town for GIs, The Bunyan Pub.

When we walked in, there was a live rock and roll band playing "The House of the Rising Sun," for a standing-room-only crowd. My standing room was about ten feet from the lead singer, whom I found fascinating. I formed an impression then that hasn't changed despite the 50 years and 50 countries I've experienced since. And that is ... that there can be found nowhere on earth, an uglier human being than an ugly Englishman.

Oh, no doubt the English have produced some outstanding beauties, but such beauties are found amongst all peoples. When it comes to ugliness, however, the English are unmatched, and this young troubadour was the epitome. The unfortunate lad was so ugly that he deserved study.

The long boniness of his skinny head and the giant gape of his huge tooth filled mouth were incredible to look at. I was nearly overcome by the impulse to walk up to him so that I could peer down his great throat when he opened it so fully to let out his raspy, ready-to-crack voice.

His head was supported on a long thin neck with the most prominent and motile Adam's apple I've ever seen. His pasty white complexion was decorated with countless pimples, and huge freckles were splattered all over his face. His ears were long and narrow and lopped over at the top. No two hairs on his red head seemed to grow in the same direction. Everything about the poor fellow was ugly.

I am unhappy to report that I cannot hear the song he was singing without being haunted by that incredibly ugly face and gigantic toothy mouth.

If only I could re-associate the song with some more aesthetic peak experience. Perhaps I could find that lady from so long ago, replay the evening, and arrange it so that when she turned on the radio "The House of the Rising Sun" would play. Then, perhaps, I could enjoy the song.

But no, my wife wouldn't approve, and such experiences can't be arranged anyway. Let me just resign myself, for the rest of my days, and perhaps beyond, "The House of the Rising Sun" and that ugly singing face as one shall be.

40. *Irish Exclamations*

In 1966 I made a hitchhiking tour of Ireland, and the most vivid memories I have of that trip are bound up in various exclamations heard from people I met along the way.

Ted, a Belfast truck driver, for example. He gave me a lift to Dublin. A bolt, holding one side of the front bumper of his dump truck, broke during our trip. The banging and clanging as the bumper hit the pavement brought us to an abrupt halt.

We searched for something to fix the problem, but all we could find was a length of rope attached to the back of the truck.

Unfortunately, it was so tightly knotted that we couldn't undo it.

We had no cutting tool, so I laid the free end of the rope over the edge of the truck bed, and with a few blows from a stone cut off enough to secure the bumper.

"Blimey, leave it to a Yank," exclaimed the astonished Ted.

He dropped me off in the center of Dublin where I heard a paper seller shouting, "Extra, Extra, read all about it." Now that's a line straight out of the movies I'd seen as a child, and he's the only real paper boy I've ever heard use it.

This fellow had a memorable face as well as a memorable exclamation. He looked as though he had gotten too close to a grenade during the war. A surgeon had patched his face back together, but each patch appeared to have a different level of circulation. The colors of the patches ranged from bright pink to white to dark gray. A patch-quilt face, he had.

The next exclamation to impress me was one of disbelief:

"Americans!" This was a student from Trinity College speaking.

The comment was addressed to me. We were in a pub, and he asked me if I liked Dostoevsky. I thought he was offering to buy me some sort of Irish whiskey.

"No," I replied, "I don't drink hard stuff, but I like Guinness."

The student took me down the street to a secondhand book seller and bought me a two volume *History of Ireland*, which he had also discovered I knew nothing about, and a copy of *The Idiot*.

I read the novel, and eventually I read many others by Dostoevsky. Perhaps my degree in English has something to do with that empathetic young student's gift. He offered the books not in criticism, but in a sincere effort to open me up to something that he saw as important.

I must confess, however, that I haven't yet taken time to read the history books, but it's only been 50 years.

My last day in Ireland was beautiful. Rain during the night had made everything fresh and clean, the sky was clear and the sun bright— a perfect spring day. All of Dublin was out in that glorious Sunday morning, as I walked down O'Connell Street in search of the ferry terminal.

In front of the terminal I noticed a very pretty lady waiting at a bus stop. She had a little boy with her who was all dressed up in knickers, jacket, tie, and dress shoes. A perfect little gentleman, it would appear.

But then he stamped in a puddle just behind his mother's heels, splashing cold water up under her dress.

She squealed, and shaking the boy by the arm, exclaimed, "Sean O'Casey, what are ya doin'!"

That exclamation, I thought, marked a perfect closure to my stay in Ireland—

"Sean O'Casey, what are ya doin'!"

41. Laredo

My Air Force technical training took place near San Angelo, Texas. Many of my friends during that time went on weekend trips to the Mexican border towns to experience “life on the wild side.”

I was a fuzzy-cheeked youth of 18, but I wanted at least to have a look at such goings-on as I’d heard about, down south of the border.

One weekend I joined some friends for a day trip to Laredo. I was going along as a spectator, so I was nominated to drive the car we’d rented. I parked on the Texan side of the boarder and we walked across into Mexico.

We entered the first bar we encountered and though it was little past noon, the bar was crowded. Men sat and stood around a long table on which a young woman danced.

She was flirting with a big fellow behind whom I stood.

Suddenly, she did something quite vulgar toward this man. Not that what she was doing right along wasn’t vulgar, but she was vulgar on top of being vulgar.

The big man was not amused. He shoved her off the table. A stream of Spanish exploded out of her like a machine gun. She ran out of the bar but quickly returned, dragging a short, fat policeman.

She stood beside me, gesticulating wildly and railed against the big bully who had pushed her off the table. Then he stood up, looking as big and as fierce as a grizzly bear.

The crowd looked at the chubby little policeman, as if to say, “Now what are you going to do?”

Much to my astonishment, he suddenly grabbed me by the collar and yanked my 98 pound weakling frame out into the street. Ignoring my objections and the peals of laughter emanating from the bar, he pointed toward the border crossing, gave me a rude shove and said, “Vamoose!”

I did, though quite confused about just what had happened.

I returned to the rented car in Texas and listened to bluegrass on the radio until my friends had had enough of Laredo.

I had gotten my fill much sooner than had they.

42. The Losing of the Stone

I slip on this heavy gold ring
With a hole where the sapphire was—
How strange to wear it.
I bought it in Bangkok
After three days bartering
And many cups of tea.
I wore it in Agra, Karachi, Kabul.
Pathan tribesmen plotted to steal it
But a young student warned me.

When my daughter came for the weekend
She handed me this ring.
It felt cold and heavy in my hand—
How strange to hold it.
The sapphire fell out on my honeymoon
On a beach at Bar Harbor—
Lost in the sand, I searched in vain.

My dreams were her dreams, she told me,
And I believed because I wanted to.
When I married I took off this ring
I put it in a drawer somewhere, and time passed.
My dreams attracted her, my dreams repelled her
And now my daughter comes

To visit
Bringing this symbol of my wandering youth—
How strange to have it.

I dreamed during the demise that I died.
I awoke trembling
Tiny flowers floated on the darkness
She could not hear my terrified heart
She slept on the far side of the bed
Holding onto her fantasies
Rejecting me.

Where is he who became intoxicated
On jasmine scent and peacock cry
In the gardens of the Taj?
The fragments have coalesced
The stone has been replaced
The ring is being worn
And with the wanderlust of my youth
I am seeking out the corners
Of a new uncharted landscape.

43. *The Return*

As we traveled from the airport out into the countryside to our training site, dusk was settling on Niger. People and animals hurried homeward. In front of huts women cooked on open wood fires, sending out the delicious smells of cooking food blended with sweet wood smoke, drying bean hay filled the air with its fragrance, and the various animals, Zebu cattle, goats, sheep and camels contributed their characteristic, earthy aromas. All these, mixed with the smell of the dusty earth, contribute to the heavy velvety perfume of an African evening in the Sahel.

These unique and varied African odors have a powerfully connecting effect on my psyche: I feel immersed in nature here; peacefulness settles within my being, and with it, the harried feelings I've been experiencing since I was last in Africa melt away.

Here in this place of poverty, life feels more real, more meaningful than I ever felt it to be in high-tech Huntsville. African villagers are preoccupied with something very basic and vital—survival. They live on the edge of failing in that endeavor. A growing season with too little rain, political disruption, and the delicate balance is broken; the specter of mass starvation descends on these gentle powerless people who are contented to have life only.

Straw houses, gourd bowls and millet—millet thrashed

with a stick, winnowed in the wind, ground with wooden mortar and pestle, cooked on a three-stone fireplace over a wood fire in a clay pot. Dust-to-dust, ashes-to-ashes is a truth undeniable here, and it is wonderfully reassuring. Love of life is intensified and its basic beauty manifested. I have been seduced by Africa, and I give myself to her magic and her mysteries with joyful abandonment.

44. The Candle

During Peace Corps training we experienced what is called a mini-live-in. Trainees go out into a village and live with an African family for three days. The experience is a self-test to see how we can adjust to being alone in a village where, for example, there is no toilet and no concept of privacy.

The reality of cultural differences tends to intrude into one's fantasy ideas about life in a bush village rather quickly. Folks of this region, for example, do not think it good to sleep alone. Therefore, they supplied volunteers with sleeping friends, usually a bunch of kids. One woman slept in a hut with seven other people—wall to wall people.

My wife Shirley, a prestigious Madame (i.e., a married woman) was treated a bit differently. She got her own place.

The Chief of BAZAGUEYZE guarded her privacy, a privilege single volunteers didn't experience. The crowd was always around watching her every move, but the good Chief (MAI GARI) kept them at a distance.

Each trainee arrived in their assigned village with a sleeping pad, mosquito net, water purification tablets, a box of matches and a candle. The chief, who had heard about western ideas of privacy, served Shirley dinner in her hut. She lit her candle and the villagers gathered around her door watching the candle in wide-eyed wonderment. Shirley promised the chief's first wife the candle as a gift as soon as she had turned in for the night.

The chief closed the door to Shirley's hut so that she could eat without an audience.

After eating, Shirley explained to her hosts that she was tired and wished to sleep. She set the dishes outside, stuck the burning candle in the sand beside them, and closed the door of her grass hut.

A crowd gathered outside her hut to watch the candle, there was a great discussion about it. Then the door to Shirley's hut opened and in came the biggest bravest man in the village nervously holding the burning candle between two sticks.

"You forgot this," he said, and carefully stuck the glowing candle into the sandy floor of the hut.

Much to the crowd's astonishment, Shirley picked up the candle with her hand, carried it outside, and called for the chief's first wife.

She held the candle out toward wife #1, saying, "This is a gift to you," but the woman dared not touch it. Shirley blew out the flame, waited a bit for the hot wax to solidify, and then the woman timidly accepted the candle and a box of matches.

She took the candle to an open spot under a tree, stuck it in the sand, and lit it with a match.

The candle spread its soft glow in a wide circle where villagers young and old gathered in the African night to marvel at the magic of a candle.

45. Coup

I had just settled into reading a stack of this week's *Herald Tribunes* when what sounded like a handful of stones hit my metal roof. They did not bounce or roll, however. I rushed outside to investigate and heard explosions and gun fire from the direction of the Kennedy Bridge 1/4 mile away, the only bridge across the Niger River in this country. It is often a site of civil disturbances. A burst of machine gunfire sent me rushing to the phone to report the situation to the American Embassy. Then I called my wife, Shirley, at her friend's home.

"We have a serious disturbance taking place here; you must come home immediately," I told her. "Do you understand?"

She did not respond; the phone was dead.

I turned to the local radio frequency and though the station was surrounded by the sound of battle, with bullets whizzing, bombs bursting, tanks and armored personnel carriers dashing all about, the radio continued to broadcast the African Cup Games from South Africa. No live media coverage of events here.

When an hour had passed, and my wife had not managed to return from her friend's home, five minutes away, I began to suspect that the authorities had cut off more than just the telephones, and that we were experiencing more than just another student riot.

Shirley had found herself blocked by heavily armed, testy young soldiers at every intersection, motioning with their weapons that she must turn around and return to wherever she was coming from. Her friends managed to lead her down side streets and alleys, eventually delivering her home.

Machine gun bursts sounded throughout the neighborhood as she entered our gate, and in the distance, toward the Presidential Palace, howitzers and cannon blasted away at government buildings.

The only source we had for information about these events was the radio. The local station was now playing martial music.

We turned one radio to the VOA and the other to the BBC, but for hours neither made any mention of events is Niger.

Finally, at 7:30, the BBC reported that all communications to the country of Niger had been cut off. They speculated that a coup might be taking place.

Suddenly I remembered that at 7:30 we were supposed to be gathering with friends at a popular restaurant, Les Roniers, for a surprise birthday party for our friend Randy. We all were getting a surprise party, I reflected, and Randy would never forget the fireworks on the day he turned 50.

Our guardian rapped at the door to inform us that there was a curfew in place and that the soldiers were clubbing anyone they found on the streets.

Finally, at ten o'clock the VOA verified our suspicions.

The democratically elected President of Niger had been arrested and deposed by the military, the constitution of the least developed country in the world had been suspended, and all political parties had been banned.

Shirley and I opened our windows and shutters and listened into the night air. All was quiet, all was peaceful, but we knew that we had experienced what folks with international careers fear most—the infamous *coup d'état.*

46. Cameroon
17 August 1989
(A Page from my Journal)

This evening while I was enjoying a bottle of pineapple pop at Nybou's, I had an experience which made me aware of how much the busyness of our minds robs us of the richness of life.

My head was full of unorganized thoughts about various projects and plans, completely oblivious to my environment.

Across the street was a high mud wall brightly illuminated by the setting sun. Suddenly, I noticed a phenomenon which jolted me awake, so to speak. A woman with a bundle of firewood on her head was passing by. Her shadow on the wall was upside down.

In trying to understand what I was seeing, I found myself becoming increasingly aware of the world around me. In front of the wall was a wide sheet of water. I knew it to be putrid sewer water, but the lowering evening sun had turned its surface iridescent gold. The sun's angle cast the shadows of passersby onto the water, and the water, like a mirror, projected the shimmering shadows onto the wall, bottom side up.

While I passed time enjoying the parade of people and animals moving along the wall on their heads, my mind was tempted to get busy with thoughts about Plato's analogy of reflections on a cave wall or to point out my observations to others, but I resisted that in favor of savoring the simple sensuous pleasures of the moment without defiling them by intellectualizing.

A duckling started to poke in the mud at the edge of the water, sending arrhythmic ripples out over its surface. Now the figures did a grotesque dance as they floated along the wall.

My enhanced sensitivity began to take in more and more of the richness around me. Nearby, the recorded music of a traditional gourd "guitar" and troubadour blasted the atmosphere as a circle of onlookers commented excitedly on the moves of two men playing an indigenous game with neem seeds.

I began to marvel at the unique taste of my pineapple drink. The cool clean glistening sand on which I sat, sparkled with the day's departing sunlight. The plethora of odors arriving from all directions on the shifting evening breezes produced an olfactory orgy, as the sights, sounds, colors, and dynamics of the adjacent market overwhelmed my every sense.

The world had been transformed into something wondrous. But you know, this miraculous richness and beauty of life is with us at every moment and in every place if we will but take the time and make the effort to open ourselves to it.

47. *Escape from Freetown*

The Revolutionary United Front rebels, called the RUF, were converging on Freetown, Sierra Leone. I notified my employer, the International Rescue Committee, that I was evacuating. Medicine Sans Frontiere arranged for expats to evacuate from Hastings Airport, a bit east of Freetown. However, the day before our departure the RUF overran that airport. Lungi Airport was across the bay, requiring a ferryboat trip. That night the RUF blew up the ferry and sank it. The next morning Medicine Sans Frontiere called, instructing me to gather my staff on Lumley Beach where there was a helicopter pad.

A Russian MI-18 helicopter flew us across the bay to the airport. Customs rushed us through, just stamping our passports without inspecting our bags. As we gathered in a glass-fronted room, a strange aircraft landed, a Russian IL-24. Its propellers did not stop as we ran to mount the rickety steps for access to the aircraft. As soon as the door closed, we took off and shortly looked down at the beautiful green sea. We were out of Sierra Leone—everyone cheered.

Then, what appeared to be smoke filled the ceiling area of the cabin. Passengers exchanged worried glances. A man came out of the cockpit, and speaking in a thick Russian accent, introduced himself as the aircraft's owner and assured us that we were not seeing smoke but mist from the air-conditioning.

With that, nearly every passenger reached into their carry-on and pulled out a bottle of hooch. We celebrated wildly all the way to Abidjan, Ivory Coast.

Yes, on 23 December 1998, our vintage Russian plane took off from Lungi Airport at 4:30 p.m. The RUF had overrun that airport by 5:00 p.m. We were on the last plane to depart Freetown for quite some time. Over the next few weeks the RUF went on a recreational slaughtering rampage in Freetown, killing some 5000 people.

That was a close one.

48. Dancing Cadaver

One of the last duties of a Peace Corps Volunteer is to return the Peace Corps dirt bike to the regional capital. My wife and I decided to return ours via a long circuitous route that would take in the fantastic landscape along the northern border of Cameroon and Nigeria.

The landscape there is more like a dreamscape. It was once the location of numerous volcanoes, which later cooled, the cores solidifying. Over the centuries the harsh climate of Sahelian Africa has weathered away the mountains around these cores, leaving them standing free–huge columns of black rock sticking straight up out of the undulating countryside.

The people living there are an ancestor-worshiping tribe called the Kapsiki.

While among the Kapsiki, we were invited to a traditional funeral for a very important man who was expected to have a place amongst the ancestral deities of the tribe. It was a big affair, and people traveled from the entire region to partake.

We first had to negotiate with the family for the privilege of attending the funeral. We paid about $15, which included photography rights. Our guide warned us not to give anyone any more money.

We were taken directly into the hut where the body was being prepared. As we entered, two men were tying the cadaver's jaw shut, and others were slipping blue trousers onto his emaciated legs. I was invited to photograph the process, but the idea struck me as being disrespectful.

My wife and I preferred to be outside where some 100 people were starting to dance. Their dress was sumptuous and colorful. The men danced carrying various weapons, swords, daggers, bows, sticks, warrior clubs, and axes, as well as canes and umbrellas. The women carried hoes and gourd bowls.

The musicians kneeled in a tight circle facing one another as concentric circles of people danced around them in alternating directions.

Each village arrived *en masse*, charging over the hill on the run, their battle cries echoing from the petrified volcano cores that surrounded the village.

Each arriving group added greatly to the tumult, and each added players to the musicians. The celebration became wilder, more noisy and drunken, and for me, more intimidating.

From time to time our guide would take us in to see the progress of the cadaver. The attendants had wrapped the torso in thick quilted material and strapped it on with strips of cloth, giving the body an erect posture.

They dressed it in a shirt with sleeves extending two feet beyond the hands and finally wrapped the body with bands of colorful cloth in a weaved pattern.

The emaciated body of the old man slowly took on a muscular youthful look. The head was wrapped in bands of white cloth until it was particularly oversized.

Then the head was decorated with twigs, sticks, various grasses, and the sign of the clan—a porcupine quill with a hunk of goat meat on it. The final touch was sunglasses.

I mounded a large rock in front of the concession so that

I could get a good photo of the entrance of the cadaver.

Then a very mean looking young man approached me and started yelling. The frenzied celebration was so loud that I could understand nothing, but his attitude was clearly unfriendly. I ignored him and focused my camera on the door through which the guest of honor was about to enter.

Suddenly, the young man produced a peculiar weapon, hatchet-like, with a disk for a blade, which he drew back as if to strike. I understood that clearly. I put away my camera and went to my guide for an explanation.

The young man was a brother to the deceased. He demanded more money for photography rights. I was willing, but our guide refused.

Then the cadaver entered on the shoulders of a husky youth. It sat there on his shoulders, the long sleeves of the shirt being used to balance it.

The noise, screams, drumming, and passion of the dancers intensified many fold. I could hear nothing of the haggling, so I handed the youth some money and turned my camera on the dancing cadaver.

The man carrying it motioned for the people to make way for me, I got a clear shot of this fantastic scene, and one click was all I wanted. I motioned to my wife to leave. Then I felt a hand on my shoulder. I turned to see the cadaver towering over me, his carrier thrusting an open palm under my nose. I put a coin in the hand and continued to exit. Then I was surrounded by countless screaming angry faces, all demanding pay for being photographed.

Now I understood our guide's advice about money. I was genuinely frightened and pushed aggressively through the crowd toward our bikes.

Our guide shouted that he would meet us on the road. We started our bikes and took off up over the hillside. When I got over the ridge I stopped, but my wife had not followed. I dismounted my bike and rushed to the ridge for a look.

There was my wife frantically kicking the starter of her flooded bike. A great crowd was headed straight for her, led by the dancing cadaver.

I ran down the hill, told her that my bike was up over the ridge and to get herself up there. I turned her bike down the hillside and eventually got it jump-started. I then roared up over the hill past hundreds of wildly drunken and well-armed lively men, and one dead one.

When I joined my wife, I found our guide with her. He told me not to worry, the crowd was only taking the cadaver around for one last look at his farm.

I was not convinced that their purpose was not to show the cadaver what they do to non-paying tourists, and we were a good many miles down the road before we felt safe again.

49. Transported 02 December 1995 (A Page from my Journal)

Struggling with a piece of French literature, I sit on the veranda of the GTZ hostel in Tillabéri, NW Niger. It is cool and there is a slight breeze. Dappled soft sunlight filtered by the undulating branches of surrounding neem trees dances across my page. I become conscious of a background noise, the growl of a small aircraft turning a circle around the village.

Suddenly I am transported across time and space to my deck on Bell Hill in Otisfield, Maine, a small plane is circling the church-crowned hill on its approach to Oxford Airport. The experience is so powerful, so convincing that I lift my eyes from my book in search of the fish pond and the other familiar sights of my so loved backyard so far away. No oaks or maples, no birch or spruce greet my glance, only strange trees, tough unfriendly trees, designed for this tough unfriendly environment, thorned and armor plated.

For a moment I seem to detect the scent of pine, but no, it's not there. Then a bird's song, never heard on Bell Hill, brings me fully back to Sahelian Africa.

Not such a bad place, really, but it's not home.

50. Opportunity Missed

During the mid-sixties I was in the Air Force, stationed outside of London. One of my favorite pastimes during that epoch was hitchhiking.

Now I had a colleague who wanted to travel to the Edinburgh Festival with me, but he refused to hitchhike, saying that he would sooner have the time saved on the express train to spend in Edinburgh than the money saved by hitchhiking.

I argued that we could hitchhike to Edinburgh just as fast as we could take the express train. My friend didn't accept this, so I challenged him to a race. He would take the express train from London to Edinburgh, and I would hitchhike.

On the day of the race I got as far as Leeds, halfway, by noon. It was outside of Leeds that a very attractive young captain in the Royal Women's Air Force stopped to pick me up.

She turned out to be from Llangollen in Wales, a place I knew well. We started talking about her home town, and before long we had established a certain degree of intimacy, and she was telling me some very private things about her life.

For example, two years previously at some sort of international military conference, she had met a pilot from the South African Air Force, they had exchanged addresses, had been corresponding ever since, had fallen in love through the mails, and in just two months, she was resigning her commission from the Royal Air Force and was taking off for South Africa to become this gentleman's bride.

However, she was experiencing considerable anxiety because there were certain aspects of male-female interactions that she didn't have much experience with, and she was afraid that she was going to be a disappointment to her new husband in this realm.

So, suddenly, I heard her making a suggestion that between wherever we were and Newcastle, where we were going, was a small village where she lived. She proposed dropping me off there with the key to her apartment. Then I could spend the day doing the tourist thing in this picturesque little village, and in the evening she would return for, shall we say, some experience.

Well, I was quite unprepared for this suggestion, it really blew my socks off. I can remember mumbling something about this very important meeting I had in Edinburgh and how I absolutely had to be there by four o'clock and thanks anyway but I must push on.

Well, she didn't say much after that, and when the road from the motorway to Newcastle appeared, she let me down.

It may be significant that the next car to stop was driven by a brain surgeon. He drove a Jaguar XKE, and when I told him about my race he looked at his watch, did some calculations, and said "By Jove, I think we'll make it," and we took off across the flatlands between Newcastle and Edinburgh at speeds often exceeding 120 miles per hour.

He managed to set me down in front of Waverly Station in Edinburgh at quarter to four. I rushed in, found the quay for the London Express, and was able to greet my friend as he stepped from the train.

Yes, I won that race. But I must confess, for the past 50 years, I have often doubted that it was worth it.

51. Arrowhead

I look to the horizon and see
Dashing to meet the cavalry
On painted ponies, red and white
Screaming braves dash into the fight
Where clash of steel and burst of ball
Are mingled with a bugle call.
The blood makes mud of the dust that rose
From hooves rushing to avenge their woes.

52. Complimentary Smile

October morning
Drive to work
Brilliant sky
Vibrant leaves
Acorns fall
Partridge flush
Wind-harried waters' white foam
World alive and *crisp.*
Intersection encounter
Complimentary smile
Distraction
Destruction and death
Glass burst into the sun
Steel crumpled and torn
Soft body broken
Smell of blood
Mourning
October
Morning.

53. Cowboy's Lament

As I look out on the morning sun
what d'you 'spose I see
but a gang of cowboys riding hard
and they are after me.
Now what on earth did ever I do
to deserve such a fate
but my own interests to protect
and my own money take.

It happened on the 12th of May
I was passing through Saleen
when just to quench a driving thirst
I stopped in at the canteen.
Well just inside the swinging doors
an old man played solitaire
he gave me such a friendly grin
and offered me a chair.

Well we'd but played a game or two
when joined by another pair
by the time we'd played a dozen more
why I was winning fair.
But then my luck it seemed to change
'twas then I did take note

that the two young fellows playing with us
sure looked like the old goat.

But when the light it dawned on me
that they were in cahoots
I'd lost everything I had to my name
except my alligator boots.
So I reached into my pocket so bold
pulled a pistol quick as a flash
and sticking it under the old man's ear
I scooped up all the cash.

Oh, surely you can understand
I did not have the time
to separate what was theirs
away from what was mine.
I pulled a lamp from off the wall
and crashed it on the floor
when carelessly I tossed a match
the flames blocked off the door.

And then I jumped on Desert Dust
and skedaddled left and right
and when the fire had been put out
I was nowhere in sight.
Oh, I always did love ramblin'
guess I'll ramble ever more
cause when them boys caught up with me
I left 'em lying on the floor.

As I look out on the morning sun
what d'you 'spose I see
but a gang of cowboys riding hard
and they are after me.
Now, what on earth did ever I do
to deserve such a fate
but my own interests to protect
and my own money take!

54. The Eagle

When Grandmother was a child, an eagle swooped down on the family farm to carry away a new born lamb and also succeeded in permanently seizing Grandmother's peace of mind concerning eagles. A half century later she watched my toddling self, curls bouncing, oblivious to danger, with a worried eye skyward. Grandmother's anxieties over dangers from the sky were imparted to my forming psyche. Sometimes an adult would point to a speck in the distance, a point no bigger than mosquito larva or a tree toad tadpole. "See the eagle," Uncle would say.

There was something numinous and final about this distant creature: danger held in abeyance, an ominous spirit lurking in shadows, the plague incubating in slum rats.

At University, as I learned about the precarious hold that eagles themselves had on life, the idea slowly, vaguely formed, that what my family and hometown folks feared about the eagle was what it symbolized—uncompromising wildness, the unpredictable freedom to swoop down into their well-arranged fields and patterned lives, their tight world of puritan propriety, as ordered, as organized and as neat as a New England cemetery.

But this insight was dim, or perhaps threatening even to me, not strong enough to wrench my life from well-worn paths, giving form and direction to my unfolding self.

Though I sought an education with an international career in mind, my spouse grew increasingly opposed to such a lifestyle. I tried the hometown domesticated route instead, but the compromise left my heart unsettled. When, from my garden or woodpile, I spied a distant speck in the sky, the illusive eagle, I would hunger after him.

My vague feelings of fear became admiration, an envious longing for the vast world, the freedom of the eagle.

Suddenly I found myself single again. In a breach of pain my world had unexpectedly reopened. I began looking for new places in it, places of mystery, magic and unpredictability.

One night, after I had been offered a job in a far-flung country, I was visited by a powerful dream. I met a huge eagle who offered me a ride on his back. His wondrous wings unfolded to slowly lift us skyward, and once aloft in the pristine air, he spread them wide to effortlessly glide.

We soared across the Atlantic, across North Africa, the Mid East, and as we entered India, I directed him toward Agra. When I saw the Taj Mahal below, I urged the eagle to alight, but he refused. We circled the site several times, me pleading all the while to be let down. The eagle turned his head and looked at me with disapproval in his eyes. Then like a rocket, we streaked back toward North America where we crashed in a fiery explosion.

I awoke from that dream, and as sharp as that eagle's glance, the understanding pierced me—the time had not yet come for me to follow the example of the eagle. I continued my adjustments to the single life amongst friends and family, in fields and on roads of my home turf.

Eventually I had mastered my new lifestyle, enjoyed it thoroughly, and met a lady who truly shared my dreams.

We married, liquidated our assets, and launched international careers.

On the day I signed the contract for the sale of my house, I returned there to bid the place a final farewell. It had been my haven for a decade, representing many parts of myself. To walk away from it into the unknown was frightening.

I opened the glass doors leading to the deck and stepped out to embrace, for the last time, the unique beauty, the spirit of that beloved space—fruit trees and fish pond, fields and forests that I loved, roads and creeks intimately known, roots eight generations deep.

With tender affection I absorbed the rich ambience of my first world, accepting its central point in my being. Then I stretched, and as I did I looked skyward at the circle of blue fringed by the tall trees around the house and yard, and just at that moment, a magnificent eagle burst into view, he was just at the tree tops. I could make our every detail of his miraculous form.

He gave me a sidelong glance before disappearing behind the trees. His glance held no disapproval, and in his sure grip he carried away any lingering doubt or fear. A sublime peace settled on me, and at that moment my spirit took flight to glide on the high winds and to seek out the corners of my very own way of being.

55. Fibber

Fibber McGee had good intention
But what he said was mostly invention.
He didn't set out to tell a lie
But from the truth his story did fly.

Perhaps he just found life too boring,
So his mind sent the facts a-soaring.
The truth he sometimes found quite smothering;
To drab gray things he'd add some coloring.

A toad became a brontosaurus;
He saw one, truly, he'd assure us.
His stalking panther was just a cat;
Fibber's mind made it more than that.

A flee became a kangaroo;
Through Fibber's eyes it grew and grew.
Fibber McGee was really not crazy,
But his keen mind was in no way lazy.

You could say he exaggerated;
For sure he never underinflated.
It wasn't fun to talk of a mouse
Unless it was as big as a house.

(3)

And when her hand it touched on me
I was all filled with fire
And leaving all her beastie crew
With her I did retire.

Her granite walls did echo loud
Her pleasure deep and dire
As eagerly by candle light
I fed my strange desires.

She woke me at the break of day
And bade me quickly fly.
I questioned not her warning tone
I dared not ask her why.

O Lorelei, my Lorelei,
What have you done to me?
Before your song disturbed my sleep
My simple life was dear.

(4)

How gaily through the wood I strove
No vine impaired my way.
When by the oak the wood did clear
The sun made bright my day.

I found my book all laid with dew
And set it in the sun
But when the sun had dried it out
The pages all were one.

I put my book upon a shelf
And there it still does lie
But locked within its pages is
The song of Lorelei.

O Lorelei, sweet Lorelei,
What have you done to me?
Before your song disturbed my sleep
My simple life was dear.

58. Little Miss Mousey

Little Miss Mousey
Lives in our housey
Munches on millet and grain.
The food of our chickens
Provides her main pickins
But we count it now as a drain.

Tail, it's quite swooshy
Curly and bushy
Gives her an air of delight.
Her eyes are most muggy
Most misty most buggy
You'd think that her corset was tight.

Likes she to chatter
Eats from a platter
Sleeps in an anchovy crate.
We would like some wee ones
Bediapered fleed ones
But first we must find her a mate.

59. Passing Through

Not the high road nor the low
But blessed none-the-less.
Blessed by people
Blessed by places, experience.
Never knowing personal want
But viewing it.
Watching the innocent,
Victims of recreational slaughter in Freetown,
Mass graves in the Balkans,
The world's indifference in Afghanistan.
People wanting only a simple sustainable life,
Caught in events
Created by the greed of others.
They pay the supreme price
While the others are welcomed on red carpets.

I, blessed with family & security,
Enough to live well,
Walk through the devastation,
The results of man's inhumanity to man,
Touched in my heart,
Shocked in my mind,
An observer passing through,
Outside it all in its midst,

For some reason blessed.
A blessed burden,
To carry through my days,
Along with joyful impressions—images
Glorious cultural variations of people,
places, beliefs, myths, music and food,
Foreign but rich,
Informing my eclectic view
Of all and everything.
Tragedy and suffering have I seen,
But grateful am I still
For the path
That has unfolded before me.

60. The Smudging

A dark dreary shadow
hovers about
From halls and stairways
its presence to shout.
Lovers & neighbors
want it to go,
A Shaman is called
her smoke to blow.

Negative ions
spread all about,
Giving remaining
blocks a stout clout.
Where will this banished
Jekyll now Hyde?
Far from our presence
may it abide.

Be gone forever
resentments & fears,
A new day is dawning
greeted by cheers.
Sacred again
this space will be,
Healing protection
now sets all free.

61. Will O' the Wisp

On the dark hill, lightly glowing
Small spots of light I see showing
My mind excited for to see
Fear trumped by curiosity
I hiked the dark lanes through the wood
Climbed the hill as fast as I could
And there amongst great boulders high
Some sort of music, sort of cry
Little people wildly dancing
'Round a fire so nimbly prancing
Sinewy bodies one foot tall
Grotesque faces, warts and all
In a circle, twisting, leaping
Fiddler, flutist their time keeping
All their voices loudly crowing
Fire-lit faces strangely glowing.

What thing is this my eyes do see
Will people ever believe me?

I yearn to join the dancing crowd
My stifled voice would sing out loud
But this scene could not include me
So slipping away I did flee

I could not tell folks what I saw
They'd think my psyche had a flaw
Some things in life but few can see
Some things we see must be let be.

62. *Breakdown in Afghanistan*

In 1966 I met a Peace Corps Volunteer in Kabul, Afghanistan. He was part of a team that studied river flow in various valleys across the country to identify potential sites for hydroelectric dams.

On one occasion they headed out toward Bamiyan with their paraphernalia in two Bedford trucks. Far out in the isolated boonies the driveshaft on one of the trucks broke. Half of the team stayed with the truck as the rest of the team in the functional truck drove to the next town, 45 miles west, to seek a new driveshaft.

A mechanic in the next village told us that he could have a used Bedford driveshaft for us by noon the next day. We took a hotel room and waited.

The next morning, back out in the desert, the stranded volunteers observed a small caravan approaching. Much to their astonishment, one camel had a driveshaft strapped on its back. The volunteers dickered with the Afghans and finally got the price down to about $200, which the volunteers had amongst themselves.

How delighted they were when the driveshaft was affixed and worked. They happily drove to the town where their friends awaited them. Entering town they saw the second Bedford parked by a teahouse. Inside, they gleefully greeted we who were to be their rescuers. "Can you believe it," they chortled, "some nomads showed up with a Bedford driveshaft!"

"We can believe it," we assured them. "They stole it from our truck last night."

63. *British Brogans*

Mountain climbing was a youthful passion of mine. Back when my knees were up to it, I found standing on a mountain top exhilarating. The mountain didn't have to be a famous peak, and I was quite contented to take the easiest trail to the summit.

Back then I don't believe folks had today's obsession with professional gear and fashion. Spandex and designer boots were unknown.

I did, however, have a special pair of boots. They were British Commando boots, which I bought at an Army Surplus store in Bedford, England, where I was stationed in the Air Force. They weren't particularly comfortable, but they were built to last forever. They had immortal soles, you might say, which after ten years of hiking showed little wear.

These soles were thick with deep tread, ideal for mounting stone outcroppings, worn smooth by centuries of erosion. I first wore them climbing Mount Snowdon in Wales and discovered that the tread tended to pick up small stones. I polished the leather after my hike, but I decided to keep the pebbles of Mt Snowdon under foot.

Next, I climbed Ben Nevis in Scotland and added her pebbles to my footwear. On top of Ben Nevis I discovered a plaque placed there by the Boy Scouts of New Mexico. It was mounted on a black rock, which the Scouts had toted all the way from the summit of Old Baldy in their home state. Since I was soon being transferred to West Texas, I set my sights on Old Baldy for my next climb. There I picked up an additional pebble or two and probably dropped some from my previous climbs.

I wore these boots hiking the Rockies in Colorado, the Appalachians in Western Tennessee, and then the White Mountains of my native New England.

Later I was assigned to Germany, which gave me access to the Alps and the Pyrenees. My stone clogged boots began to take on a special charm for me. They had randomly exchanged bits of mountains from all over the world, and I took secret pleasure in the knowledge of it.

Next, I found myself in Pakistan, where I hiked the Hindu Kush along the Afghan border, followed by the ascent of Mount Dopesar on the other side of Pakistan in the Swat Valley. This is not an internationally famous mountain, nor is it particularly high, but it is one of the Himalayas.

After viewing the world from atop a Himalayan peak, I returned to my Pakistani host's home in Peshawar. There I found myself contemplating my bruised brogans with their collection of pebbles. I fell into a kind of trance before them as images of my many hiking exploits flooded my mind, adventures tangibly represented by those bits of mountain jammed in the tread. These boots had charisma.

I later joined friends at Green's Hotel for tea and cheese cake and talked much of this mystical experience with my beloved boots.

Then I returned to my host's house where I was greeted by his bearer, Bashrat.

"Sahib," he said, giving me a mostly toothless grin, "I

65. *Sahib*

During the 60s, when I was teaching English in Peshawar, Pakistan. I preferred a bicycle as my mode of transportation to the crowded public busses where I might be expected to share a seat with a goat or some chickens.

Along the route to work I passed through the edge of a concession containing about twenty adobe huts. Some days I traveled during the early afternoon, siesta time, when all sensible adults were sleeping. In fact, I became known as "that foreigner who rides his bicycle during siesta."

Children in this part of the world are strictly disciplined, especially concerning respect toward elders. But like children everywhere: when the cat's away, the mice do play. During siesta, when this peculiar looking foreigner passed by, there were no adults awake to observe what happened.

The children decided to play with this fellow—me. They started throwing clods of dirt and stones as I passed by, calling me *foreingee,* foreigner. I knew their conduct was not acceptable by local standards. I also knew that a stone in the hands of one of these children was a weapon. I'd seen them pick a crow out of flight with a hand-thrown stone. Their projectiles were accurately controlled not to come too close to my head, but close enough to worry me.

I was irritated by this frequent occurrence, but decided to ignore it. Then one day things went too far.

As I approached the collection of huts, one boy sat on a bullock cart at the edge of the road. Just as I passed him he hawked. I thought, *You'd better not,* then he spit, just missing the back of my head. I was furious!

I turned my bike down the alley between two rows of huts, clanging my bicycle bell. Confused men jumped to their feet from grass mats on the ground. Soon I was surrounded by some twenty people.

Pointing at the frightened lad, I demanded, "Show me the father of that boy."

Soon, Armadou, blurry-eyed, stumbled out of one of the huts. Someone had dragged the boy in from the road by the collar. "Is this your son?" I demanded of Armadou?

"Yes, Sahib," he replied, bowing politely.

"This boy and others throw stones at me when I pass here, and today your son spit at me. Is this the way you have taught your children to respect strangers?"

The man silently reached through the crowd and yanked his son by the shirt into the center of the assembly. He then began to beat him, with open hand and closed. My first impulse was to stop the beating, but I knew that this was the way such matters were handled here.

The boy moaned some, but never cried out; his father beat him to the ground and added a kick for good measure.

The crowd looked satisfied that justice had been done. Though I was consumed with conflicting emotions, I managed to calmly shake the father's hand, the chief's, and several other local dignitaries' hands, thanked them for their cooperation, and departed.

From that day on, whenever I passed this little village,

whatever the hour, the children rushed to the edge of the road, saluted, and called me Sahib, honored elder, as I passed by. And no one saluted more smartly than did the young lad who had suffered such a beating for his rudeness.

66. Socks of Sand

In the 60s I lived in Peshawar, Pakistan, with a professor of English and his family. They were upper middle-class and could afford an automobile and one servant, a young man of about 20, named Bishna.

Their youngest son returned to America with me and eventually became an American citizen. Fifteen years later I returned to Peshawar with him to partake in his wedding. The fortunes of the family had changed over the fifteen years. Thanks largely to the considerable business success of two of their sons, the family was by this time decidedly upper class, with an army of servants.

The original servant, Bishna, was now head bearer, as they are called in the sub-continent. He wielded his authority with pride and sometimes a heavy hand. The family had just hired a new bearer, a young lad of twelve. He was instructed to be totally at my service. Although his English skills were quite basic, he was eager to please, and we managed to communicate.

Just prior to my trip I had undergone arthroscopic surgery on my right knee. and my doctor had given me a series of exercises to do which involved lifting weights with my legs. Weights were improvised from knee length stockings filled with sand.

Naturally, I did not carry socks of sand to Pakistan with

me, but only a pair of rugged nylon knee length stockings to be filled with Pakistani sand. A few days after I had settled in, I noticed that across the street a new building was being constructed, and in front of it was a pile of sand.

I called the young bearer, Shezad, and asked him to put some of this sand into my stockings and to deposit them in my room. Meanwhile, I departed for the post office to send off a clutch of postcards.

When I returned I detected that something was wrong. The young servant, Shezad, looked particularly unhappy, and quickly made himself scarce. Bishna looked angry, and the entire family had an uncomfortable air about them.

It was tea time and we were served in the garden. Delicately, my host broached a question.

"Henry, what is it you have asked my servant to do with your stockings?"

"I asked him to fill them with sand," I replied.

"Why did you want sand in your stockings?" he continued.

I explained the use of the sand filled socks.

Orders in Pushtu were quickly issued to the gardener, who ran out of the gate and up along the outside of the garden wall.

It seems that when Bishna saw the new servant enter the house with my socks full of sand he was convinced that the young lad could not have understood me correctly. Certainly no one in his right mind would want a perfectly good new pair of stockings filled with sand. He gave the lad a thrashing for being so stupid and for ruining the stockings of an honored guest. He then threw them over the garden wall for emphasis.

The servant sent to retrieve them had no luck. On the

other side of the wall was a tent put up by a family of Afghani refugee squatters. When the socks flew over the wall they had assumed that the rich family on the other side was throwing them gifts. They had already emptied them of the sand, washed them out, and had sent a son off to the market to sell them.

More orders were given in the local patois, this time to the chauffeur, who drove off in a hurry. Before we had finished the second pot of tea, he had returned with a new pair of socks. They were knee length and brown, like the originals, but they were of cotton, not nylon.

The head bearer, Bishna, looked considerably uncomfortable, but the reprieved Shazad appeared to be most happy when he was dispatched to the construction site to fill socks with sand again. I tied them together at the top as all the household gathered to see what strange things this foreigner wanted to do with sand filled socks.

I draped them over my ankles to demonstrate my leg lifts. Unfortunately, the seams in the socks split completely open and all the sand spilled out. Awkwardness again settled on the scene. I quickly retrieved two pair of rugged socks from my room and sent Shazad off again to the sand pile. I tied them all together and demonstrated my leg lifts without incident. The family and servants of my host were then happy that feelings of embarrassment were no longer necessary, but happiest of all was the junior servant, Shazad.

67. Ribbons and Heart

During the 60s I was in the Air Force stationed on a communications base about two hours north of London. The operations center was located on a hill to the back of the base about a half mile walk from the chow hall. The base offered a shuttle service between the chow hall and operations, but I preferred to walk.

Once I was headed up the hill after lunch when the passing shuttle bus hit a bird which landed, stunned, in the gutter nearby. When I tried to pick it up the bird ran along the gutter, and finding a drainage grate, it squeezed itself between the bars and disappeared.

When I looked down the drain hole I saw the bird floundering in deep water. Surely, it would drown. I tugged at the grating over the drain, but it wouldn't budge. Then I wiggled it here and there, and though I loosened it some, I could not lift it out of its hole.

"What are you doing, Airman?" an authoritative voice pronounced.

I looked up to see the base commander standing before me, the single star of a brigadier general glistening from each shoulder.

I jumped to attention, saluted and then explained that an injured bird had fallen into the drainage system and that I was trying to rescue it. I felt that I had been caught at something foolish. I assumed that a general would find such consideration of a bird to be a rather silly concern for an "American Fighting Man."

"Let me help you," he said, kneeling on one side of the grate. We removed the drain cover and I retrieved the bird.

I put the bedraggled fowl on the sidewalk, and it ran off into the tall grass.

"No broken bones," I said. "She'll be okay."

We replaced the grid, stood and saluted. I thanked him and we parted ways. The good general now had a round grease spot on one knee of his otherwise immaculate uniform, all for saving a drowning bird. Would he share the story during lunch at the officers club, I wondered? Would he let out what I had discovered: that at least one general had a beating heart under all the rows of colorful ribbons?

68. Territorial Imperative

During my wandering youth I once traveled to Amsterdam. I was reading Robert Ardrey's *The Territorial Imperative* at the time. This book deals with the expression of dominance among animals.

An animal, it seems, expresses dominance by stance, glance, or by a slight movement toward a subordinate beast. I was thinking about these things as I was crossing a long wooden footbridge that led to the Vondel Park in Amsterdam.

A flock of seven ducks swimming upstream passed directly under me. I struck an aggressive stance, as if to say, "Don't you ever swim under MY bridge again." Immediately the ducks started to swim in nervous circles, eager to get back under the bridge. They swam toward the far bank of the waterway and reaching the furthest possible point from me, they turned to sneak back under the bridge. I took one step in their direction, and they immediately turned around, quacking their disapproval of my self-imposed authority.

Even though the bridge was a good hundred feet across, I had managed to establish dominance over them by a mere gesture.

Soon a group of ducks came *down*stream. I stood where they would have to pass under me and put my hands on my hips. The ducks stopped in confusion and seemed to be discussing the situation. They decide that they had misinterpreted my posture and continued on with apparent confidence. I took a sudden step toward the railing and leered over it at them.

The whole flock turned as one and splashed back up stream in a panic, where they were joined and consoled by the first group.

About 15 ducks looked up at me with that one-eyed duck look and quacked complaints amongst themselves about the injustice of it all.

I had the bridge to myself for the better part of an hour.

By the end of that time I had countless water fowl backed up from the bridge: ducks, geese, swans and mud hens. I would let fowl parties cross under the bridge from *down*stream, but once they were on the upper side of the bridge, defied their return.

The greater the web-footed population grew, the more daring and frantic they became and the more dominant I had to be. The aggression escalated from assertive stance to bold step, stamped foot, flung arms, and finally to square footed jumping.

I was about to resort to shouting when I saw a policeman approaching.

"May I ask what you are doing?" he demanded.

"Well, er, ah," I explained, as ducks by the dozens made a splashy dash under the bridge, "I'm doing an experiment on animal behavior."

I then produced Ardrey's book and gave a quick synopsis of how animals accomplish dominance and suggested that we try it on the approaching band of geese.

"I'm afraid you'll have to find another place to do your experiment," said the policeman. "You see, you've upset some ladies in the retirement home across the bridge."

I looked to see a five-story brick building with old ladies scowling at me from nearly every window.

"They consider these ducks to be their pets," continued the policeman. "They have called in a complaint that you are harassing them."

I gave the ladies a friendly smile and waved, but none of them returned my friendliness.

I promised the policeman that I would do no more experiments on the ducks of Amsterdam and was allowed to continue on my way.

I do recommend Ardrey's fascinating book to anyone interested in animal behavior. I can vouch for the validity of his thesis, but I do not recommend testing it in public gardens.

69. Ants

One of the exciting events of my childhood summers was the ant wars. My father bought a plot of land at the edge of the woods and built us a home there. We soon discovered that our side yard was the site of a yearly confrontation between the red army and the black army.

Suddenly in early August they would appear: the red army from the woods, and the black army from under an ancient pine tree by the road. They would battle day and night for several days, and then they would disappear, leaving only their dead as testimony to what had taken place there, an ant Antietam.

I'm not an entomologist—never studied the subject—but I do find God's wee, antennaed creatures fascinating.

In my varied travels I have encountered numerous interesting variations on the theme of the ant. In Afghanistan I discovered an ant with great long legs, reminiscent of a spider, though with six legs instead of eight. Their bodies hung suspended from their knees. They ambled along like most ants until they were startled. Then they extended themselves to their full height and dashed off at astonishing speed, on their tippy-toes as it were. I found it humorous to observe and took pleasure in startling these creatures into a stampede.

In the African rain forests I've marveled at the machinations of army ants by the millions, coursing across jungle floors in their sunken roads. There were soldiers who maintained the roads, perhaps sergeants, and there were the masses of grunts, privates, scurrying with huge burdens to some mysterious point.

The most fascinating of the army ants were the traffic cops—lieutenants, maybe. They were at least four-times larger than the road-crew ants, which were twice the size of the grunts. They would stand at strategic points in characteristic poses, their huge heads thrown back, pivoting left then right, their menacing mandibles held open.

"Turn left, turn right, steep incline ahead."

What were they communicating? I could only imagine, unable to hear their commands or to interpret their body language. Their fellow ants, however, understood clearly.

In Cameroon I encountered a very practical ant. It was a small variety, the kind that makes little volcanoes on roadsides and in school yards. But these ants were particularly efficient; they didn't close the circle of their mounds but left one side of the excavated debris open, so that during their goings and comings they didn't have to ascend and descend their mound. They could march straight to or from their hole in the ground.

Very practical ants, these, but not very bright.

If you tapped the ground near one, it would turn around and run back to its hole, following a well-trodden path. But if you knocked one a few inches from its path, it would run in panicked circles and could not find its way home.

Overly regimented ants, I'd say.

Years ago I bought 25 wooded acres in rural Maine and set about building a home there. Beside the house clearing I discovered a huge wood ant mound, nearly as big as a Volkswagen.

Wood ants have a highly developed symbiotic relationship with nature. They collect decaying matter and put it in underground chambers where they use it to grow mushrooms and various fungi to dine on.

This colony used wood moss to their advantage, as well. Moss grew on their mound and held it together. When expanding or heightening their mound, they were cautious not to bring up so much earth that they smothered the moss. They would bring up about half an inch and then would wait for the moss to adjust to the new condition. This was the secret to the massiveness of their mound.

Everyone attracted by my house building activities, the excavators, the contractors, the carpenters, plumbers, landscapers, neighbors and relatives were without exception eager to advise on how to get rid of these ants.

"I'll just live in peace with them," I'd say, eliciting raised eyebrows.

Some took my wife aside and warned that the ants would destroy our house, ruin our food. Her husband's peculiar attitude was incomprehensible to them.

We finally made a compromise, she and I, that if ever she found one of these wood ants in our house, these easily-identifiable, medium-sized, shiny black ants, that I would destroy their entire empire. But first we would behave more neighborly toward them.

My new human neighbors began to suspect that an eccentric was moving into town. What convinced them was when I put a fence around my ant hill. Yes, a fence.

You see, whenever guests came with children, the little ones could not resist running upon the attractive moss covered mound. Their feet would sink deep into it, and their little shoes would fill with unhappy ants. A scene would follow.

Also, my friends, the ants, would have a home subject to erosion, because the protective moss would have been damaged.

Yup, I built a protective fence around this ancient ant heap—to protect the children, of course.

Now, certain ants did come into our country home, every April, when the last traces of snow had melted and the earth thawed out.

My wife was not happy about this. She feared dire results—the house would crumble, the freezer would be emptied.

"Show me—what harm are they doing?" I would ask.

Just being there was harm enough, and no argument could assuage her revulsion.

But not all ants are the same, you see. I will not share my house with pismire ants—so named for their smell—mainly because they enjoy pies as much as I do. I will not share my raspberry pie with a pack of pismires.

Carpenter ants, their name a contradiction: carpenters build houses, these creatures tear houses down—no, I will not have them tearing mine down.

But these ants in our house were a wee, tiny sort, which I suspect were hanging out in our nice dry house until the high spring water table sank a bit.

I tried to convince my wife that they had come to help with spring cleaning. They carried off those minuscule toast crumbs from corners and cracks. They were never found in fresh food, they preferred decaying debris, stuff we didn't want around anyway.

But no, a chemical-biological battle would ensue. My wife sprinkled magical powders around the house foundation, killing Lord only knows what, and numerous tiny tins, resembling mini restaurants, fast food joints of diabolical intent, were placed around in strategic locations.

And then there was the spoon treatment. Whap! Whap! Whap! Tiny ants rendered to spots of oil on the counter, disturbing my weekly ritual with Saint Paul Sunday Morning on Public Radio.

The battle usually raged for two weeks, then an announcement would be made, Pentagon press conference style—VICTORY ACCOMPLISHED—the ant army enemy had been vanquished.

One spring my wife left. She never returned, but the ants did, in April, right on schedule. I ignored them, though I did consider paying them a few cubes of sugar as restitution for the evil done to them in the past.

After two weeks, when the water table had sunk a few inches, or the last toast crumb had been found, or whatever, they withdrew.

To the last departing straggler I called, "Thanks for the help, see y'all next spring now."

No, it is not that I love ants, necessarily; it is just that I try to get along with all God's creatures. They have their purpose, and I have mine. Let us live harmoniously.

70. *The Drama of Tracks in the Snow*

When I stepped out into a world freshly decorated with a coating of sticky snow this morning, I noticed evidence that an animal had sped through the backyard taking great leaps, fifteen feet between each landing.

"Why would a deer run through our yard like that?" I wondered?

On closer observation I was astonished to see that they were rabbit tracks, huge ones that only our Giant Flemish, Bun Buster, could have made.

Then I noticed two sets of tracks approaching his, diagonally from the woods—coyotes.

The rabbit tracks originated from where Bun Buster had bedded down for the night by the compost pile. It appears that the coyotes surprised him in his cozy bed.

Bun Buster took off, leaping a great arc through our backyard while the coyotes, a pair, transversed his route. Within the first 100 feet of the chase, they were on him, then he pulled quick maneuvers to outwit them, snow and forest floor kicked all about from their "hare" turns. One of Bun Buster's maneuvers brought him crashing into the pasture fence; they scuffled, but he escaped under the chicken house. He was safe!

Unfortunately, in his panic, he didn't feel safe, and he made a great miscalculation—he ran out from under the other side and into the field. The coyotes were on him again, cutting didoes, spinning, slipping, falling in the frantic fatal dance of predator and prey.

Again Bun Buster slammed into the fence, and a coyote closed in to grab him. He escaped again but the second coyote nailed him against the fence a few yards further on.

The biggest coyote dragged their 18 pound trophy across the field and into the woods. I followed the path clearly marked by the dragged carcass for about 150 yards. The path entered a grove of small hemlock. I circled it and saw coyote tracks exiting the grove but no longer dragging our beloved rabbit.

I got on my hands and knees and crawled through the tight thick branches. I found a mound of snow and leaves with a familiar pair of large back feet sticking out—Bun Buster, stored for a later meal!

For nearly two years Bun Buster has delighted us. He was an old cull from a rabbit farm. When I brought him home I set him down in our backyard. Our geese did not welcome him. They gathered around him, hissing and honking, but Bun Buster was an unflappable beast. He mostly ignored them. They soon became friends.

His life had previously been spent in the one square yard of a rabbit breeder cage. Suddenly, he had the whole world to explore, but he seldom wandered over 100 feet from where I first set him down.

Bun Buster was social. He liked hanging out with the geese. Whenever we appeared he came lippity-lopping up, head lifted to be scratched, and hoping for the treat of a piece of bread. We supplemented his wild diet with a dish of pellets on our porch. He often ate and lounged there as we enjoyed our supper, watching him through the glass door.

And now our friend is gone. He will no longer delight and entertain us. We knew the freedom we gave him put him at risk, but isn't freedom worth risk? Caged, he would have remained living, but free, his violent death ended a real life.

71. Butterfly

Years ago I took a job in Maine delivering for UPS. The fellow responsible for training me was very much a product of the military. He marched, he didn't walk, he carried his neck bolt-upright like a marine—it helps them not to think, I'm told—and he barked his commands like a drill sergeant. In fact, his nickname was, Sarge.

We were delivering at a small island off Boothbay Harbor. Southport Island, the home of Rachael Carson, is accessible by bridge.

The island is ringed by a tar road, and dirt roads crisscross through its center. We were taking a shortcut on one of these dirt roads when I came upon a low spot where a large water puddle had gathered from a rain storm.

The water had mostly evaporated by that time, but there was a wide ring of highly moist mud around the puddle's edge which had attracted well over 100 monarch butterflies.

I hit my brakes and slid to a stop, the butterflies all became airborne, and I slowly drove through their swarm.

"My (expletive deleted) word!" barked Sarge. "Don't tell me you brake for butterflies!"

"What did you expect me to do, run over all those beautiful creatures? Have you no reverence for life?" I replied.

Sarge threw his head back and banged his forehead with the heel of his hand several times. Once he had regained his composure, he calmly looked at me and asked, "Henry, do you think you're really cut out for this job?"

He didn't understand, reverence for butterflies goes deep in my personal history. We in my family have a special place in our hearts for those delightful spots—collections—of color, dancing over fields of flowers. One can forgive God for mosquitoes when one considers butterflies.

There is a favorite family story about Grandmother and butterflies. When her youngest daughter, Constance, went to high school, her biology teacher assigned the class the responsibility of collecting butterflies for the lab. He instructed them to capture the insect, to pin it to a pine board, and then to drop iodine on its head to kill it.

Aunt Connie caught a yellow butterfly, pinned it to a board on the kitchen table, and just as she was about to drop the fatal iodine, Grandmother entered the room.

Grandmother was appalled by what she saw, the crucified butterfly was unpinned and released, Aunt Connie got a paddling and was grounded for two weeks. The biology teacher got his ears pinned back, as they say in Maine, by Grandmother's tongue. No child of hers would be permitted to participate in cruel practices against such a beautiful and innocent creature as a butterfly.

So you see, my affinity for butterflies has deep roots, and yes, that UPS grunt was correct. I was not cut out for that job.

72. Mr. Pheasant

The Maine Fish and Game Department experimented with the introduction of ring-necked pheasants during the early 1950s.

By the end of the decade the Chinese bird proved to be incapable of adapting to the harsh environment offered by the state of Maine. Unlike the northern grouse, the pheasant did not know how to burrow into the deep snow to find food. The ring-necked pheasant would never be established in Maine.

Many people much enjoyed hunting these big colorful birds, however, so the state of Maine continued to raise them for release, only to be shot by hunters or to perish in the Maine winter.

I've never approved of this practice, so when 15 pheasants were released in a field near my home I designed a trap and managed to capture two cocks and a hen. I put them in my poultry house and promised them release in the spring.

One cock died and the daughter of a guest opened the henhouse door, allowing the hen pheasant to escape. Only Mr. Pheasant, as we called him, remained.

The winter was ferocious that year. I often asked Mr. Pheasant if he were not happy to be in a cozy hen house with all the cracked corn and poultry mash he wanted instead of being out there in the blizzards. Though he never responded, I suspect that he found keeping company with chickens to be a bit offensive.

Mr. Pheasant's health remained high, but as winter waned his beautiful plumage suffered degradation. His long tail feathers were broken off, feathers here and there stuck out of place, and his feathery coat was full of litter dust. In March he suffered the ultimate humiliation when he made the mistake of standing under the chicken roost at the wrong time. His natural beauty and brilliant color became so degraded that my heart went out to him. The snow was melting, however, and soon Mr. Pheasant would be released into the lush springtime.

One morning quite early when the rising sun bathed the front of my henhouse with golden light, I was surprised to find a dainty hen pheasant sitting on the outside window ledge. Her bright shining feathers were a symphony of browns, and every one was perfect. On the inside window ledge, separated from her by 1/4 inch of glass, sat the rumpled Mr. Pheasant.

The hen jumped from the window ledge and scooted into the snowbound bows of a small pine tree.

I entered the hen house and picked up the pitiful Mr. Pheasant, took him outside, and set him on the snow. He found his suddenly widened world intimidating at first, but eventually, with hesitating step, he proceeded toward the pine boughs from where eager eyes watched until his shabby form was enveloped.

I completed my poultry house chores and went back outside to see a pair of pheasant tracks leading across the snow from the little pine into the woods beyond.

I then understood that, at least for a ring-necked pheasant, it is better to suffer the indifferent vagaries of nature than to suffer the secure humiliation of a hen house.

I hoped that Mr. Pheasant would forgive me for the indignities I had put him through and that soon his plumage would match the perfect beauty of his newly found lady.

73. Rambo the Rambunctious Ram

Today started at 6 a.m. with a phone call. Yesterday I delivered our freezer lamb to the butcher for "transformation." The butcher called, apologizing for the early hour. The problem was that when the butcher house staff arrived this morning they found about half of the sheep out of their enclosure, wandering about the shed. The animals had knocked off or eaten the papers attached to each cage identifying owner, animal, and cuts desired.

"Which cage was your animal in?" they wanted to know.

"The first one."

"How might we identify your animal?" they asked.

"He has a magnificent head with great horns, about 175 pounds and still sexually intact." (unusual for a freezer lamb).

"Oh, I know which one he is," said the caller.

Later in the day when I went to pick up my animal's hide, the crew there had analyzed the problem. The door hitch on my ram's cage was bent straight from forcing the gate out; the other broken hitches had been forced in.

There were lots of lady lambs scheduled for slaughter this morning. Being the biggest and only sexually intact male in the place, he had a jolly good time with all the ladies he had managed to break out from their confinement. His final hours were much enjoyed. Good for him.

74. Zihada

I have a handsome flock of Brahma hens. These birds look so proud and stately in their milky white plumage, with its black tipped shackle and saddle feathers, black tail feathers, and feathered feet. They were the favored poultry of our New England forefathers. Because of their small comb and wattles and their feathered feet, they could better withstand New England winters than could other breeds, even though they originate in India.

I have one rooster and six hens. One hen is rather aloof for her species. She is always off by herself somewhere. While visiting India, I noticed this sort of isolation between Hindu and Muslim members of the population, so I have decided that my hen must have been converted from Brahman Hinduism to Islam; therefore, I have given her the Islamic name, Zihada.

If the rest of the flock is in the garden, Zihada is in the woods scratching up old leaves. If the flock is at the pond for a drink, she may be found in the driveway, taking a sand bath.

This style of living involves greater risk than does the style of her co-hens. She is the one most likely to be stalked by the fox, coyote, or bobcat.

But her style has its advantages too. For example, this morning I tossed an overripe tomato into the backyard for my flock's consumption. Over competition for sole possession of this colorful treat, the tomato was torn open and strewn about. Finally, one old hen got hold of it and took off on the run, bits of the tomato leaving a trail behind. The rest of the flock was in hot pursuit, all but Zihada, that is. She wandered up to where the original skirmish had taken place and calmly picked up the scattered pieces.

Then she followed the trail of strewn tomato debris across the lawn to the sunflower patch, where the second skirmish had taken place. There she had a second feast in the cool shade of the giant flowers while Sita, pursued by Ramdas and the rest of his harem, dashed noisily around the fish pond, the tomato getting smaller all the time.

By the time the tomato was no bigger than what one swallow could accommodate, happy Zihada had calmly consumed all of the tomato she wanted. The rest of the exhausted flock had known only the frustration of trying to keep the tomato from the others, without having a moment to enjoy the process or the tomato.

I am learning from my hen Zihada.

75. Seeking Normalcy

My last international assignment was for United Methodist Committee on Relief. The purpose of this project was to help war refugees return to their farms and to reset their lives back to the normality they had once known. Below is my summation of that experience.

Russian helicopter gunships roared over the Shamali Plain, north of Kabul, Afghanistan, and targeted the village of Rabat. In one of the houses, 11-year-old Ghulum Rabani huddled with his family as bombs exploded outside. Then Rabani's greatest fear happened, a bomb hit his adobe home directly, blowing him to unconsciousness.

Every member of Rabani's family was gravely injured. His father died from injuries, but the others recovered. Rabani lost one leg. Eventually his mother fled with her children to a refugee camp in Peshawar, Pakistan. Sixteen years after the Russian gunships had destroyed so much of his life, Rabani is back in Rabat. The Russians have long ago withdrawn in defeat. The puppets they left behind have met violent ends, and the scourge of the Taliban has come and gone.

It was the Taliban and their scorch-the-earth approach to the Tajik-dominated Shamali Plain that has made sustainable living impossible across Rabani's native province. Their plan was not only to drive out the supposed supporters of the Northern Alliance but also to make refugee return impossible. The Taliban destroyed shelters and all means of economic livelihood. Nut and fruit trees were cut down, and vineyards were burnt. The irrigation system was ruined, and tens of thousands of landmines were buried in fields that once produced enough wheat to feed the community.

The repressive structure of the Taliban collapsed quickly once the West brought its power to bear in Afghanistan. Proud Afghans, suffering under the indignity of refugee status, were eager to return to their native villages. What they found on their return, however, was deeply discouraging—how could life in such ruins be possible?

I first visited Rabat as the United Methodist Committee on Relief Nongovernmental Organization (UMCOR NGO) Head of Mission, in June 2002. I encountered anger from some returnees. A family head named Wahid said to me: "We returned because we heard promises from the rich countries that our country would be rebuilt, but what we find is destroyed homes, without water or other means of support. We can't live here. If we don't get help soon, we'll have to return to refugee camps. There, at least, we had shelter and food, our children were educated, and we had access to medical clinics."

Fortunately for Wahid's family, as well as for the majority of the families in Rabat's associated hamlets, UMCOR was formulating a plan to make their return to the area sustainable. Applying its participatory, multi-sector, integrated approach, UMCOR was about to implement a sustainable return project through its local partner, Rural Rehabilitation Association for Afghanistan (RRAA), which would allow Wahid's family to reestablish a meaningful life back on its ancestral land. RRAA is a project developed by Norwegian Church Aid in Pakistan.

Further, as news of UMCOR's plans reached Rabat's exiled people in Kashmir, hundreds of neighboring families were encouraged to return to the hamlets of Rabat. By the end of 2002, 250 of the 350 Rabat families had returned and are confident that they now have what is necessary to rebuild a meaningful life there.

Recipients Work Hard

The project is participatory in that it is based on the expressed needs of the beneficiaries. UMCOR's partner, RRAA, interviewed returnees to the village and traveled to the refugee camps in Kashmir, where most of the Rabati people had fled, to find out what they needed in order to return and stay.

The villagers prioritized the minimum requirements for sustainable return: shelter, water, agricultural supplies, and access to credit. UMCOR and RRAA set about creating a project that would meet these needs.

Within six months, 200 homes for the most needy families who agreed to return were constructed through UMCOR's self-help approach. The beneficiaries were required to make their own mud bricks and to build their houses up to the roof. UMCOR supplied lintels (door beams), roof beams and other roofing material, doors and windows, and a bag of cement. For water, villagers were required to dig down to the water table, averaging 45 feet, at which point UMCOR would drill an additional 75 to 100 feet. UMCOR also supplied a diesel engine, pipes, and a pump for each well. Of the 35 slated wells, 20 were pumping water by the end of 2002, averaging 5700 gallons per hour to a total of 151 farming families. Beneficiaries of these and other aids were selected by representative village councils called shuras.

A Multi-Sector Approach

Two hundred farmers were supplied with improved wheat seed, fertilizer, and pesticide. Winter wheat is now turning the recently demined fields of Rabat green again after years of no productivity. Next spring, UMCOR will distribute rain-fed wheat packages, fruit and nut tree saplings, and grape root stock. Microcredit has been thus far extended to 29 widows and 22 other individuals who were disabled by war injuries.

The vast majority of those who were injured were not combatants but civilian victims of gunships, mines, stray bullets, or shrapnel. They, like Rabani, can now support themselves with dignity as tailors, shopkeepers, bicycle mechanics, and animal raisers.

A visually impaired man named Shawali was particularly creative. Since there are no tractors in Rabat, he saw an opportunity to raise steers. He had only 60 percent of the required capital, but with a $200 UMCOR loan, he was able to purchase a pair of traction animals. Though he is unable to use them himself, he rents them with his handmade plow to neighboring farmers for $6 per day during the fall and spring planting seasons. He will repay his loan easily. Others, like the shopkeepers Ambrlaw and Murad, earn a little over one dollar profit per day, while the tailor Rabani averages between $1.50 and $2.00 per day. Though minimal, this income permits sustainability.

A Long-Held Tradition

As for the widows, we knew that they would prefer credit for a cow or two sheep, but we could not interview them in this traditional rural society. They are still behind the high mud walls around their homes and only venture forth covered in blue or white. They lift their burqas for ease of vision in the immediate vicinity of their homes or remove them to work in the fields. But with the approach of any stranger or vehicle, these tent-like garments drop down over heads and bodies, covering all.

Despite tales to the contrary, this has been a rural custom for many generations. I traveled in Afghanistan in the mid-1960s, before 80 percent of the Afghans alive today were born. The only female faces I saw in public were foreigners. The Communist takeover in the 1980s brought a lifting of the burqa and the unprecedented flow of women into education and public service. This long-established cultural practice of wearing the burqa was common for peasant women, but the Taliban brutally forced the practice on the educated, Westernized women of the cities as well.

So today, over one year since the fall of Taliban rule, provincial women still honor the burqa, not because of a government edict but out of a cultural imperative. Though I have visited Rabat regularly for seven months, I have not yet seen any of the 500 faces of the women who must be living there. Only female staff members from RRAA have had any contact with the women of Rabat.

A Community Reunited

The village mullah (religious and community leader) says that everyone in Rabat has benefited directly or indirectly from the UMCOR project. Most important for him was having the opportunity to end the misery of exile and the separation of families. During the war, the mullah escaped to the north while his wife and children fled to Pakistan. When the family was reunited after two years, Mullah Basir learned that his eldest son had died during the family's separation. Not to have been able to support his wife or to bury his son in Rabat is a great source of grief to him still. The pain is now mitigated somewhat by seeing his community reunited, shelters reconstructed, and water flowing onto wheat fields freed from landmines. His broken heart is mending, he says, and he is happy to be back home.

Mullah Basir wants the United Methodist churches that contributed to this project to know that their generosity is profoundly appreciated.

Little Mochtaba is watching his father, Hadal Shah, put the final touches on their new home. He has put mud around the newly arrived window frames to secure them tightly into the adobe wall. UMCOR's partner, RRAA, is in the process of delivering window glass to those who have their windows properly installed. With the window glass in place, Hadal, his wife, and Mochtaba will then move out of his grandfather's crowded house into their own ample space.

Hadal married four years ago, two weeks before the Taliban took over Rabat and drove most residents into exile. He returned in June to find his vineyard dead, with no water for his field, and no shelter. Now Hadal's family has shelter, five acres of wheat planted and irrigated, and a hopeful future.

A Strong Resistance

Abdullah Jhan joined the resistance against the Russians as a young man. One evening as he stood guard from a rooftop over Rabat, a sniper's bullet tore through his left leg. That was in 1986. His life as a fighter was over, but he survived, recovered, and returned to Rabat. There, though on crutches, he was able to work in the family vineyard, leaving the eventual defeat of Soviet forces to his colleagues in arms.

Generally, all ethnic groups fought the Communist forces, but after their final defeat, no single leader emerged to reunite Afghans. The various warlords brought civil war to Afghanistan, much supported and encouraged by foreign powers. Eventually the Taliban brought a certain order to the war-ravaged country as they expanded from their first victory in Kandahar. But when they spread north of Kabul, they found stiff resistance from the Northern Alliance, led by the Tajik, Masood.

When the Taliban attacked Masood's stronghold in the Panjshair Valley, they had to pass through the Shamali Plain, an area populated mostly by Tajiks. As the Pushtoon Taliban attacked Masood in the north, many Shamali Tajiks attacked from the rear. The Taliban and their defeated fighters fled from the Northern Alliance supporters to their haven in the south.

Flee or Die

When the Taliban regrouped to invade the Panjshair Valley again, they took no chances with the Tajiks of Shamali. They delivered an ultimatum that the entire population must leave the country, or be exterminated. The remaining population of Rabat and surrounding communities fled, most to Pakistan. Abdullah Jhan was among them.

Once the population had fled, the Taliban ravaged the Shamali Plain, destroying everything that would make the return of refugees possible.

After four years as a refugee, Abdullah is now back home. Thanks to UMCOR, his family has shelter and access to irrigation water. Disabled in the war, Abdullah qualified for an UMCOR Income Generation loan. With it he has purchased a milking cow. Native cows of this region are hardly bigger than a full-grown ram. During the winter months, they produce only a quart of milk per day, but when the spring rains turn the earth green with vegetation, he expects to get a gallon of milk per day from his mini-cow, a market value of 60 cents.

Sami Khan and Abdul Moqsood felt too old to go into exile and stayed in Rabat as long as possible. But with the 1999 Taliban ultimatum, survival there became impossible and they joined their neighbors in Kashmir. Rabat was abandoned. These proud men have bitter memories of being exploited in Kashmir, where they could not speak the language, were treated with disrespect, and where, in old age, they had to labor from dawn to dusk just to earn enough money to survive. Rabat for them, with all its destruction and limitations, is paradise compared to life as a refugee. They express deep satisfaction and gratitude for the UMCOR program that is making possible a return to normal life where they belong.

A Will to Survive

Sher Mohammed also fled in the 1999 mass exodus. This fiery proud man detested his life as a dependent refugee. Twenty days after the fall of the Taliban, he was back in Rabat energetically cleaning up the rubble of his home. On the third day of his return, Sher stepped on a landmine buried in his yard. Neighbors rushed him to the Red Cross hospital in Kabul. Sher lost his left leg and eye and all the fingers of his left hand.

Three months later, Sher walked out of the hospital on his new artificial leg. When, in August 2002, news reached Rabat that UMCOR had approved a multi-sector reconstruction project for the community, Sher sought reassurance from the community council that he would be included in the beneficiary list for the UMCOR shelter and microcredit. Assured, he sprang into action. Borrowing money to hire help, he pulverized the rubble of his adobe home with hammers and recast it into mudbrick.

When UMCOR visited Rabat shortly after project approval, I was astonished to find Sher's house construction well under way. There in the middle of the activity was Sher, standing deep in mud, his pants rolled up exposing his artificial leg, and his fingerless left hand forming balls of mud to fling to other workers as they formed the mud foundation of his future home. Around his yard were a thousand bricks drying in the sun. Sher's home was one of the first completed under the UMCOR project.

Sher's fortitude and resolve is not uncommon in Afghanistan, where, after 23 years of war, this nation's stubborn survivors return to their ruined villages in unprecedented numbers. More than in any other country where UMCOR has worked, the Afghans show the drive and the work ethic to rebuild their lives, economy, and nation. May the donor nations give them the basic support they need and deserve.

About the Author

Henry Hamilton is a retired international aid and development professional who spent nearly a quarter century working in some of the most dispossessed and war-torn countries in the world.

His impetus for this lifestyle developed from his childhood access to a closet of National Geographic magazines, dating from the publication's earliest days. They excited his imagination to seek a career that would enable him to experience exotic lands the world around.

Henry found a way through various international organizations including the Peace Corps, GTZ (a German equivalent to USAID), the International Rescue Committee, the Organization for Cooperation and Security in Europe, the United Methodist Committee on Relief, and others.

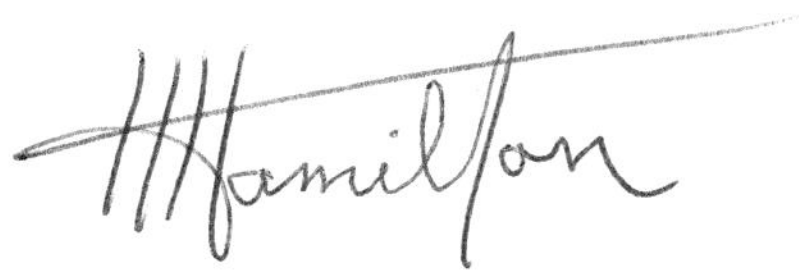